MOONS

LOST
THE
PLOT

A Lost the Plot book, first published in 2024 by Pantera Press Pty Limited
www.PanteraPress.com

Please send all permission queries to:
Pantera Press, P.O. Box 1989, Neutral Bay, NSW, Australia 2089 or info@PanteraPress.com

A Cataloguing-in-Publication entry for this book is available from the
National Library of Australia.

ISBN 9780645817997 (Paperback)

Cover and internal design: Elysia Clapin
Publisher: Martin Green
Editor: Camha Pham
Proofreader: Katherine James
Fact checker: Karla Pincott
Printed and bound in China by Shenzhen Jinhao Colour Printing

Pantera Press policy is to use papers that are natural, renewable and recyclable products made from wood grown in sustainable forests. The logging and manufacturing processes are expected to conform to the environmental regulations of the country of origin.

MOONS

The Mysteries and Marvels of our Solar System

KATE HOWELLS

FULL MOON ABOVE POPOVEC.
CREDIT: NEVEN KRCMAREK.

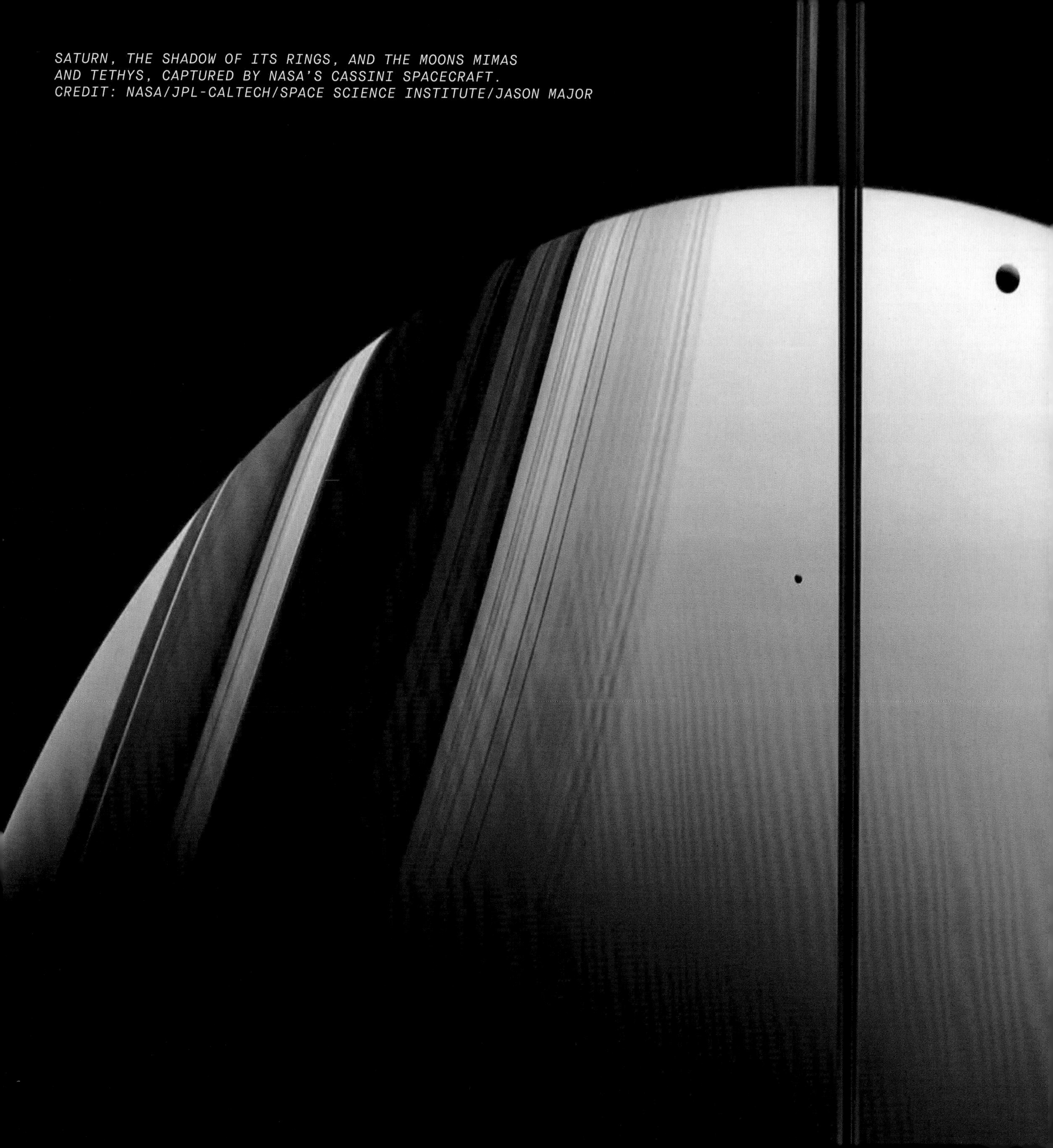

SATURN, THE SHADOW OF ITS RINGS, AND THE MOONS MIMAS AND TETHYS, CAPTURED BY NASA'S CASSINI SPACECRAFT. CREDIT: NASA/JPL-CALTECH/SPACE SCIENCE INSTITUTE/JASON MAJOR

Contents

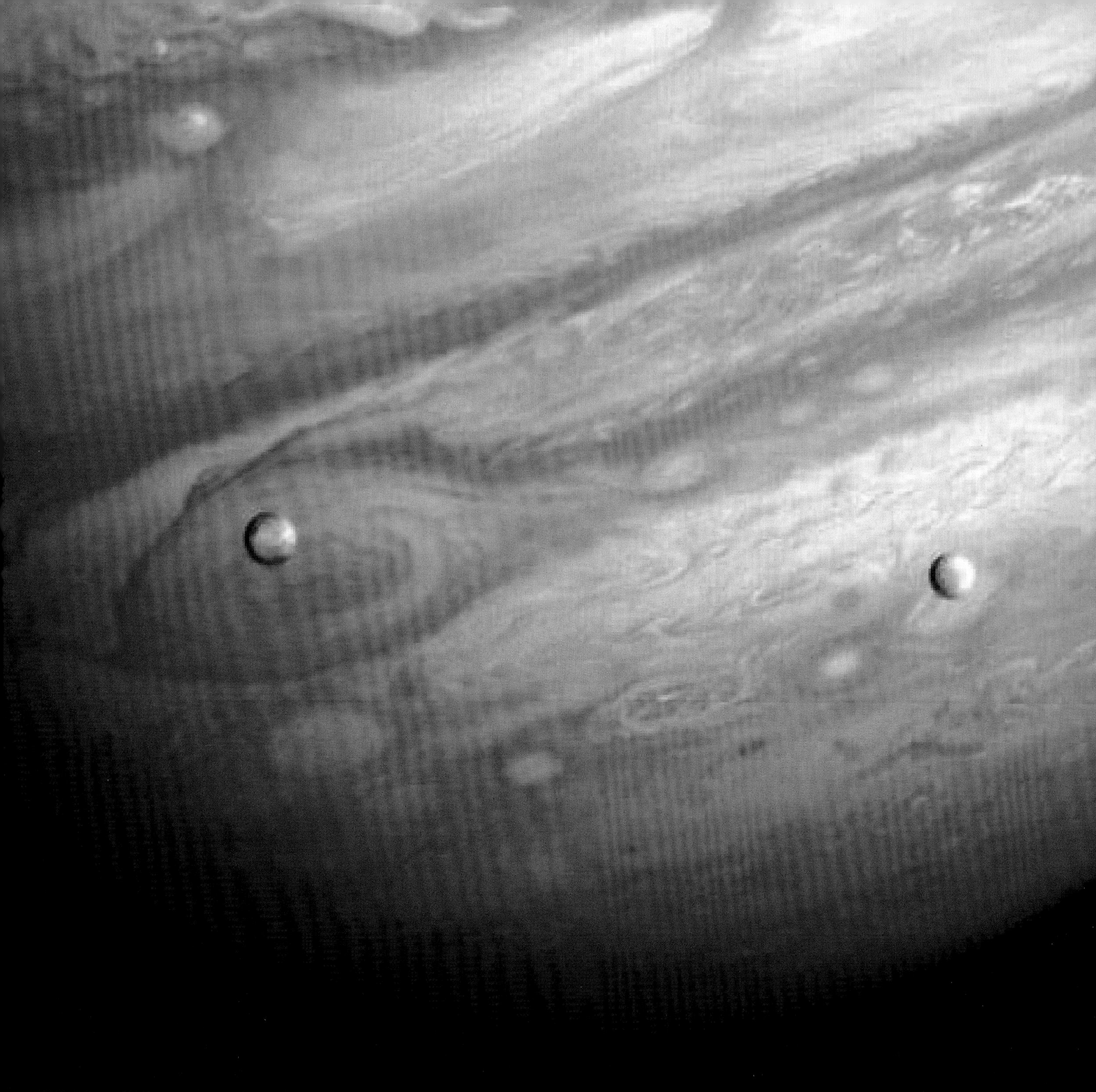

JUPITER WITH ITS MOONS IO (LEFT) AND EUROPA (RIGHT), IMAGED BY THE VOYAGER 1 SPACECRAFT. CREDIT: NASA/JPL.

INTRODUCTION

At some point in elementary school, everyone learns the names of the planets.

You might have memorised a mnemonic like 'My Very Energetic Mother Jumps Skateboards Under Nana's Patio' when Pluto was still a planet, or 'Mustard Volcanoes Erupt Meaty Juicy Sandwiches Up North' once it was demoted.

You probably learned key facts about each one: Jupiter is the biggest, Mercury is the smallest (post-Pluto demotion), and Saturn has rings. You were almost certainly already familiar with the Earth's Moon, and maybe your teachers had you sketch out its phases and draw its craters. But for most students, other moons aren't on the syllabus.

A lot of people may not really know what a moon is. A famous example of this is a clip from a shopping-channel show that went viral a few years ago. The two hosts are talking about a pattern on a cardigan, and one says, 'It almost kind of looks like what the Earth looks like when you're a bazillion miles away from the planet Moon.' This nonsensical comment then sparks a conversation that exposes the hosts' total confusion about what's what in space:

'Isn't the Moon a star?'

'No, the Moon is a planet, darling.'

'The Sun is a star ... is the Moon really a planet?'

'The Moon is a planet, honey—'

'Don't look at me like that!'

'—it's a planet!'

'Is the Sun not a star?'

'I don't know what the Sun is.'

They turn to consult people off camera, who confirm that the Sun is indeed a star and the Moon is not a planet.

'The Moon is such a planet, I can't even stand it.'

'The Moon is not a planet!'

'What else is it if it's not a planet?'

The off-camera people, having googled what the Moon is, tell them that it's a moon. This doesn't help.

'I believe it's a star! Or something. Didn't you do that thing in grade school where you had to name the planets? And there was Uranus, and there was Saturn, and the one with the rings, and dadada, and then the Earth ... the Moon is never in there, dude, it's not a planet.'

The off-camera googlers follow up: the Moon is a natural satellite.

'But things live on it, that means it's a planet ...'

'The Moon is a natural satellite? That's what Google said? No, I don't like that at all, I don't know what it means!'

And they move on, back to the much safer and more familiar territory of cardigans.

This back-and-forth probably represents the slightly more uninformed side of the knowledge spectrum, but I doubt it's too far from the average. For a lot of people, the Sun is the day orb and the Moon is the night orb, and that's all they need to know. I've had conversations with people who are uncomfortable seeing the Moon in the daytime because it seems wrong. And for many people, the experience of trying to find out more about our celestial bodies can end up a lot like the shopping-channel hosts' experience.

The truth is that the vast majority of people don't need to know things about space to get by in their day-to-day lives. Plus, a lot of people's experiences with science in school leave them feeling isolated from or even intimidated by science. If you're bad at memorising scientific vocabulary, or find long, dry descriptions of the parts of the cell tedious, you can come away from science classes with the belief that science is boring, or maybe even that you're not smart enough to do it.

This, to me, is a tragedy. The process of doing science as a professional scientist may not be for everyone, but learning things about the world around us – about the weird and beautiful reality we live in – is something everyone can do.

That's why the moons of the solar system are my favourite low-hanging fruit when it comes to new, awesome things to discover in space. You don't need any kind of skill or background in science to appreciate how extraordinary it is that there's a moon with volcanoes that blast lava right out into space, and another that has massive dunes of sand that look like coffee grounds, and another that might have life swimming in an ocean underneath a kilometres-thick shell of ice. And even if you already know a bit about a few moons, or even about most of them, there's still a wealth of information to dive deeper into. I've been working in the space field for a decade, and there were still new and staggering things that I learned while researching this book. And with an array of telescopes and spacecraft out there exploring the solar system right this second, there are always new discoveries being made.

So, to start, it's worth getting one quick clarification out of the way:

WHAT EXACTLY IS A MOON?

A FULL MOON SEEN FROM THE INTERNATIONAL SPACE STATION. CREDIT: NASA.

WHAT IS A MOON?

As the googlers from the viral shopping channel show clip rightly said, 'The Moon is a moon' – that is to say, what we call 'the Moon' is just one of many moons – and a moon is a natural satellite.

A satellite is any object that orbits another object, which is why we use that same word for the spacecraft we send into space to circle the Earth and measure weather patterns or beam down TV broadcasts. A natural satellite is something not made by humans that orbits something else. Most moons orbit planets, but there are some moons that orbit dwarf planets like Pluto.

At the time of writing, more than 290 moons are known to exist in our solar system. About three-quarters of those moons orbit Jupiter and Saturn, which together have 241 known moons.

The vast majority of moons are small – less than 250 kilometres across. Many are no bigger than asteroids. There's even a subcategory of moons called moonlets for particularly tiny moons that orbit smaller objects like asteroids. But there are also a few beasts out there, including two that are larger than the planet Mercury and seven that are larger than Pluto.

This book introduces you to 19 of the moons found in our celestial neighbourhood. They were chosen by a very scientific method: my personal preferences. These are the moons I think are the coolest and most interesting. Most are on the larger size, in part because the bigger an object is, the more dynamic and interesting it's likely to be, and the more we've probably been able to learn about it. But a few of the moons in this book are little guys who are punching above their weight in terms of scientific intrigue (and occasionally, cuteness).

Just because kids in school typically learn about planets and not moons doesn't mean that the former are more interesting or important than the latter. It's my opinion that the moons of the solar system are actually more intriguing and exciting than the planets themselves. And every single one of them still holds secrets that only further exploration can uncover.

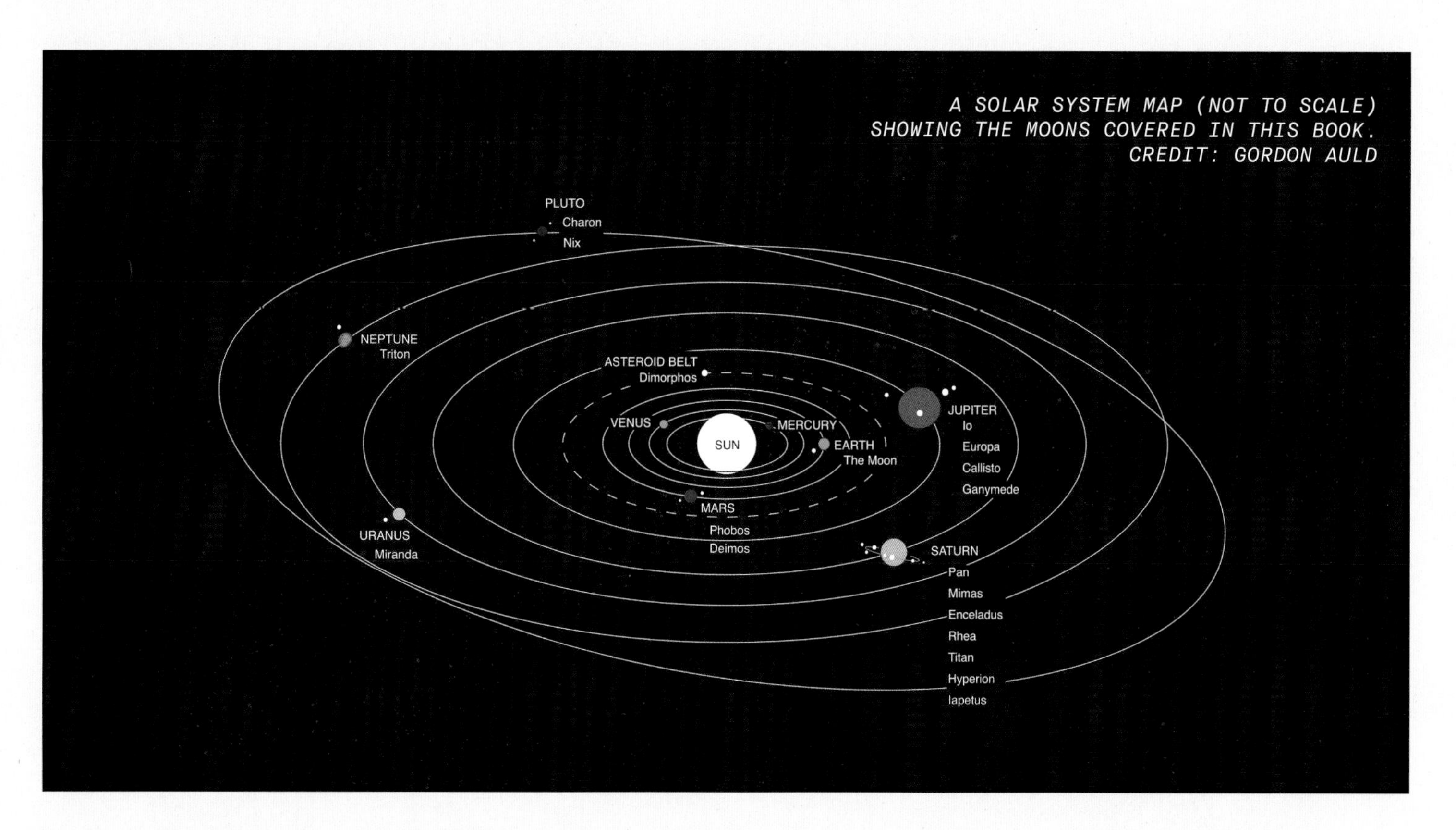

A SOLAR SYSTEM MAP (NOT TO SCALE) SHOWING THE MOONS COVERED IN THIS BOOK. CREDIT: GORDON AULD

SO, WITHOUT FURTHER ADO, LET'S MEET THE MOONS.

IO

THE VOLCANO MOON

First off the mark, we have Jupiter's moon Io.

In contrast to Earth's cool, white, stable Moon, Io looks extremely alien. It is bright yellow across most of its surface and complemented by swaths of white. It also has the planetary equivalent of acne, with spots of red, brown, orange and black.

Io also behaves very differently from our Moon. While ours is still and quiet, Io is violently active. In fact, Io is hands-down the most volcanically active place in the solar system, with almost constant eruptions happening all over its surface. There are hundreds of volcanoes, some of which spew epic amounts of lava and shoot out huge plumes of sulphur and sulphur dioxide. These plumes can sometimes be hundreds of kilometres high, shooting right out into space.

So, how exactly did this moon get so wild?

The reason comes down to Io's location. It is in orbit around the solar system's most intense and dramatic planet: Jupiter.

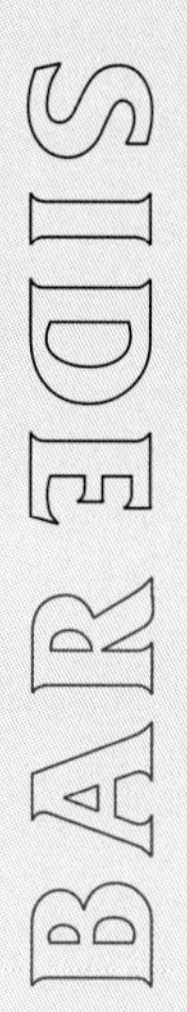

GRAVITY

Every physical thing out there in the universe (and here on Earth, for that matter) exerts a gravitational force, pulling other things towards it. The more massive something is (that is, the bigger and heavier it is), the more gravitational pull it has. Something small – like you or me or this book – doesn't have enough gravitational pull to really affect anything else. But something huge like a planet has enough gravitational pull to draw other big stuff in. That's how moons get stuck in orbit around planets: they are gravitationally pulled towards the planet and also moving in space with their own momentum, so rather than falling right into the planet they fall into orbit around it. The key takeaway: big stuff has big gravitational pull.

JUPITER'S MOON IO, IMAGED BY THE GALILEO SPACECRAFT. CREDIT: NASA/JPL/UNIVERSITY OF ARIZONA.

SOME QUICK JUPITER FACTS FOR YOU

FACT 1

Jupiter is big. It's the largest planet in the solar system – so large that 1300 planet Earths could fit inside it. In a lot of diagrams Jupiter doesn't look that much bigger than the next biggest planet, Saturn, and that's true; it's only about 15 per cent bigger by volume. But Jupiter is three times as massive. It's so huge that its gravity crushes the atoms at its core, which shrinks the planet's overall size, but not its mass. Jupiter is twice as massive as all the other planets, moons, asteroids and comets combined – including Saturn.

FACT 2

Jupiter is a gas planet. Unlike Earth, which has a solid surface with an atmosphere of gas around it, Jupiter is mostly just a big ball of gas. The 'surface' is a layer of raging storms, with wind speeds of over 600 kilometres per hour. Farther down, the pressure from all the gas on the outer layers becomes intense enough that the gases in the inner layers turn into liquid. Imagine fog getting thicker and thicker until eventually it's just water – but in this case, it's liquid hydrogen. Somewhere at Jupiter's centre there might be a solid core, but we don't know for sure.

FACT 3

Jupiter is super powerful. When the atoms at its core get crushed by its sheer hugeness, this generates heat and radiation. Jupiter itself generates almost twice as much heat as it receives from the Sun. In fact, if Jupiter were a few orders of magnitude bigger, it would start crushing atoms so intensely that it would start fusing them, which is what stars do. Basically, Jupiter is somewhere pretty far along the continuum that connects planets to stars – make a planet big enough and it becomes a star.

FACT 4

Jupiter has a lot of moons. It's got at least 95 known moons – I say 'at least' because between the time I write this and the time you read it, that number may well grow. We are finding more all the time. Four of these moons are big and important: Io, Europa, Ganymede and Callisto. Together, the four big moons are known as the Galilean moons.

SIDEBAR

GALILEO

Way back in 1610, the astronomer Galileo Galilei pointed a telescope at Jupiter and saw four dots of light whose position around it changed night after night. (You can see them too, with even a small backyard telescope.) Galileo deduced that these must be objects that orbited the planet. This went against the dominant understanding of the universe at the time, which was that everything in the cosmos revolved around the Earth. People believed that God had created the Earth and everything on it, and figured that something this important had to be the centre of it all. So when Galileo saw evidence of things revolving around something other than the Earth, it went against more than just the prevailing model of the universe; it was a challenge to God Himself. Or at least, that's how the religious leaders saw it at the time. Galileo got in huge trouble for his discovery, but he was (as we know) right. So these Galilean moons have a little place in history for helping society move away from the geocentric (Earth-centred) model of the universe.

JUPITER'S MOON IO, AS CAPTURED BY THE JUNO SPACECRAFT'S JUNOCAM. CREDIT: NASA/JPL-CALTECH/SWRI/MSSS/TED STRYK

Io is the innermost of the four Galilean moons, orbiting closest to Jupiter. Next out is Europa, then Ganymede, and then Callisto. We'll talk in more depth about those other moons, but the main thing to note here is that these moons are big. Ganymede is even bigger than the planet Mercury.

The reason this is relevant to our discussion of Io is that it's the combination of being in orbit around Jupiter and having these other large moons orbiting nearby that causes Io to be so intense.

Ultimately, it comes down to bigness. Jupiter is extremely big. The Galilean moons are also big. And through this bigness, they all have major gravitational effects on Io's orbit. As Io orbits Jupiter, sometimes it passes close to its neighbouring large moons as they go along their own orbits. As these big moons pass each other, they pull on each other slightly. This makes all of their orbits around Jupiter a bit uneven. Scientists call these uneven orbits 'eccentric', and the degree of eccentricity is how imperfect the circle of its orbit is. (This is where we get the adjective 'eccentric' that we use to describe unconventional people.) The fact that Io's orbit around Jupiter is eccentric means that it comes closer to the giant planet at some points in its orbit, and farther at others.

Gravity is stronger the closer you are to a large object, which is why astronauts float in the International Space Station – they're about 400 kilometres above the Earth's surface, and at that distance the Earth's gravitational pull is weaker. The same thing applies to Io. When it gets closer to Jupiter, the gravitational pull of the giant planet is stronger. When it's a bit farther out, it gets pulled on a bit less. And because Jupiter is so huge, this relatively small change in pull has a dramatic effect on Io.

As Io gets closer to and farther from Jupiter, the gravitational ebb and flow distorts the shape of the moon. The part that's facing Jupiter bulges up when it's closer, and recedes down when it's farther. It's kind of like how the Earth's oceans get pulled on by our Moon when it's overhead, which causes the tides (gravity in action once again). But the ocean's tides fluctuate by about 16 metres at their most extreme – and we're talking about water, not solid ground. On Io, it's the moon's rocky surface that rises and falls by about 100 metres with each 'tide'.

All that squeezing and stretching and movement of Io's rocky crust creates more or less constant friction within Io. And – as you'll know if you've ever rubbed your hands together to keep warm – friction creates energy in the form of heat. Inside Io, the heat caused by all this friction is enough to melt rock into magma.

Io's surface, which is exposed to the extreme cold of space, is solid rock. In fact, Io and Earth's Moon are the only large rocky moons in the entire solar system. Everything else is either small and rocky, or large and icy.

Beneath Io's surface, there is a massive ocean of magma. We have the same magma layer on

Earth, under the crust of rock and dirt we call the ground, but you typically don't get magma layers on smaller celestial bodies like moons.

Scientists think that Io's magma layer is about 50 kilometres deep. To put that in context, the deepest part of Earth's ocean is a little less than 11 kilometres deep.

As the moon gets squeezed over and over again, this magma jostles around. This has two major effects.

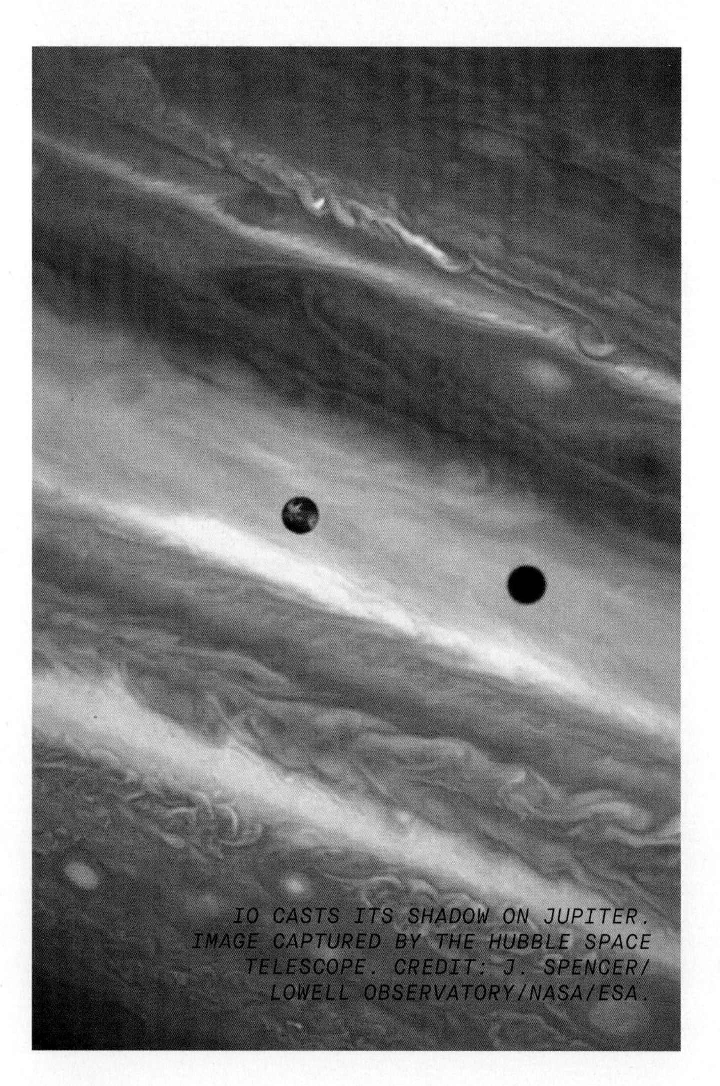
IO CASTS ITS SHADOW ON JUPITER. IMAGE CAPTURED BY THE HUBBLE SPACE TELESCOPE. CREDIT: J. SPENCER/ LOWELL OBSERVATORY/NASA/ESA.

First, it moves the solid rock above it, in the same way that Earth's tectonic plates move above our own magma layer. Like on Earth, this movement of the solid rock over liquid magma sometimes causes the solid rock to bump into other parts of solid rock, which creates mountains. This process is particularly intense on Io, so much so that it has mountains that are higher than Mount Everest. The other effect of all the magma jostling is a truly terrifying amount of volcanic activity. Almost every rocky planet in our solar system, and many moons, show evidence of a volcanic past. But today, active volcanism is super rare. Io is the only other body in the solar system that has the kind of volcanic activity we see on Earth. Instead, other worlds have cryovolcanism – a very cool phenomenon where temperatures are so low that ice acts like rock and water acts like magma. Venus shows some signs that it might still have a little volcanic activity, but it doesn't appear to be substantial. For now, Earth and Io are the only worlds that undoubtedly have red-hot liquid rock spewing out of volcanoes onto the surface.

Since Earth is about four times the size of Io, you might expect it to have more volcanic activity. But of course, Io puts Earth's volcanism to shame. Io's surface is covered in more than 400 active volcanoes, which together produce about 100 times more lava (the word for magma once it reaches the surface) than all of Earth's volcanoes combined. Io's lava is also hotter than any lava on Earth.

The largest and most powerful volcano on Io – and in the solar system – is called Loki, named after the shape-shifting god from Norse mythology. At its centre is a giant lake of lava about 200 kilometres across. This beast of a volcano erupts with impressive regularity, about every 500 days, and the eruptions are so fiery that they can be seen from telescopes on Earth, more than 600 million kilometres away.

One of Io's other most steadily active volcanoes has been producing about 100 cubic metres of lava every second for the entire time we've been keeping tabs on it. This is enough lava to overflow an Olympic-sized swimming pool every 30 seconds.

When lava flows on the surface of Io, it doesn't slow down as quickly as lava does on Earth. Because Io is so much smaller and therefore has much weaker gravity than Earth does, lava can continue to ooze along for a long time after coming out of the ground. Lava flows that stretch 500 kilometres long have been found emanating from a single volcano.

Some of Io's volcanoes shoot out plumes of sulphur and sulphur dioxide, noxious gases that stink of rotten eggs. These stinky plumes can climb as high as 500 kilometres above the surface, shooting right out into space. I'm going to insist that you pause for a second and think about what this means. Mount Everest is less than nine kilometres tall. Outer space technically starts 100 kilometres above Earth's surface. A jet of super-hot gas shooting 500 kilometres into the sky is an eye-wateringly intense thing to imagine. If this all weren't enough, these massive plumes are also moving at breakneck speeds of one kilometre per second and can gush continuously for months on end.

Together, Io's hundreds of volcanoes that constantly spew fresh lava onto the surface are doing their part to keep the moon looking young. The fresh lava covers up older rock, constantly renewing the surface. Most planetary bodies are covered in impact craters, which is an inevitable consequence of a solar system filled with random bits of rock that smash into things. Io doesn't show a single one – every crater gets filled in with lava; in fact, every single part of Io's surface is made of such fresh lava that it's still actively cooling off.

All of this is astonishing, yes. But this is only the lava – Io's volcanoes do some heroic spewing of gases as well.

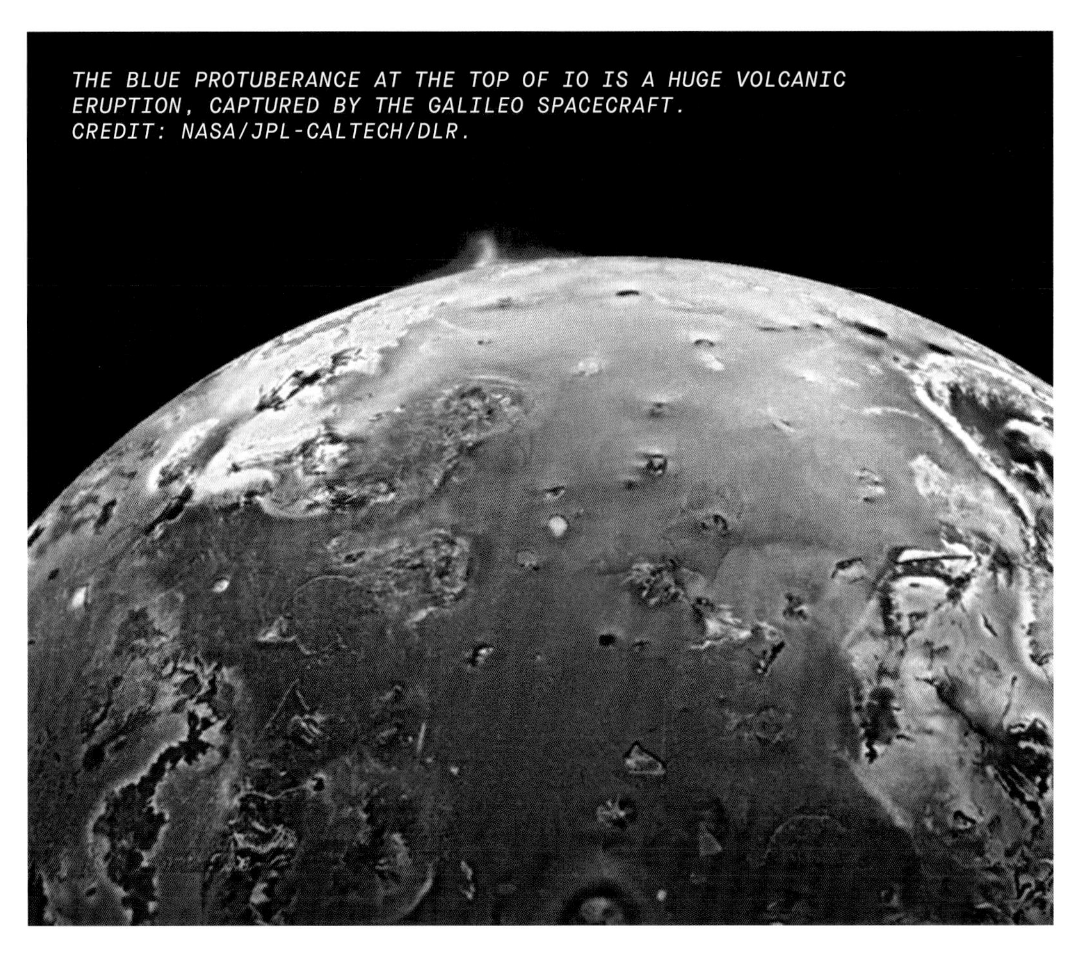

THE BLUE PROTUBERANCE AT THE TOP OF IO IS A HUGE VOLCANIC ERUPTION, CAPTURED BY THE GALILEO SPACECRAFT. CREDIT: NASA/JPL-CALTECH/DLR.

IO JUST BEGINNING TO PASS IN FRONT OF JUPITER. IMAGE TAKEN BY THE CASSINI SPACECRAFT. CREDIT: NASA/JPL-CALTECH/SSI/KEVIN M. GILL.

But there's more. The sulphur gas being shot out by Io's volcanoes hits the coldness of space, which causes some of it to harden into its solid state and fall back down onto the moon like 'metallic snow'. Some of the gases also make it all the way out into space and get swept up by Jupiter's enormous gravity. And when I say 'some of the gases', that might be an understatement to say the least – about a thousand tons of gas from Io gets added to Jupiter's heft every single second.

When these gases hit Jupiter's powerful magnetic field, they have an energetic reaction with it, creating a ring of plasma around the giant planet a bit like the aurora we see in Earth's polar skies. When NASA's Pioneer 10 and 11 spacecraft flew by Io in the 1970s, they tried to take photos of these massive auroras, but the radiation near Io was so intense that it destroyed the images as soon as they were taken.

There's hardly anything about Io that isn't scarily intense and extraordinary. You'd think it would be the coolest moon out there, but the rest of the solar system's moons give Io a good run for its money.

AN IMAGINED VIEW OF AN ASTRONAUT TAKING A CORE SAMPLE ON IO. CREDIT: GORDON AULD.

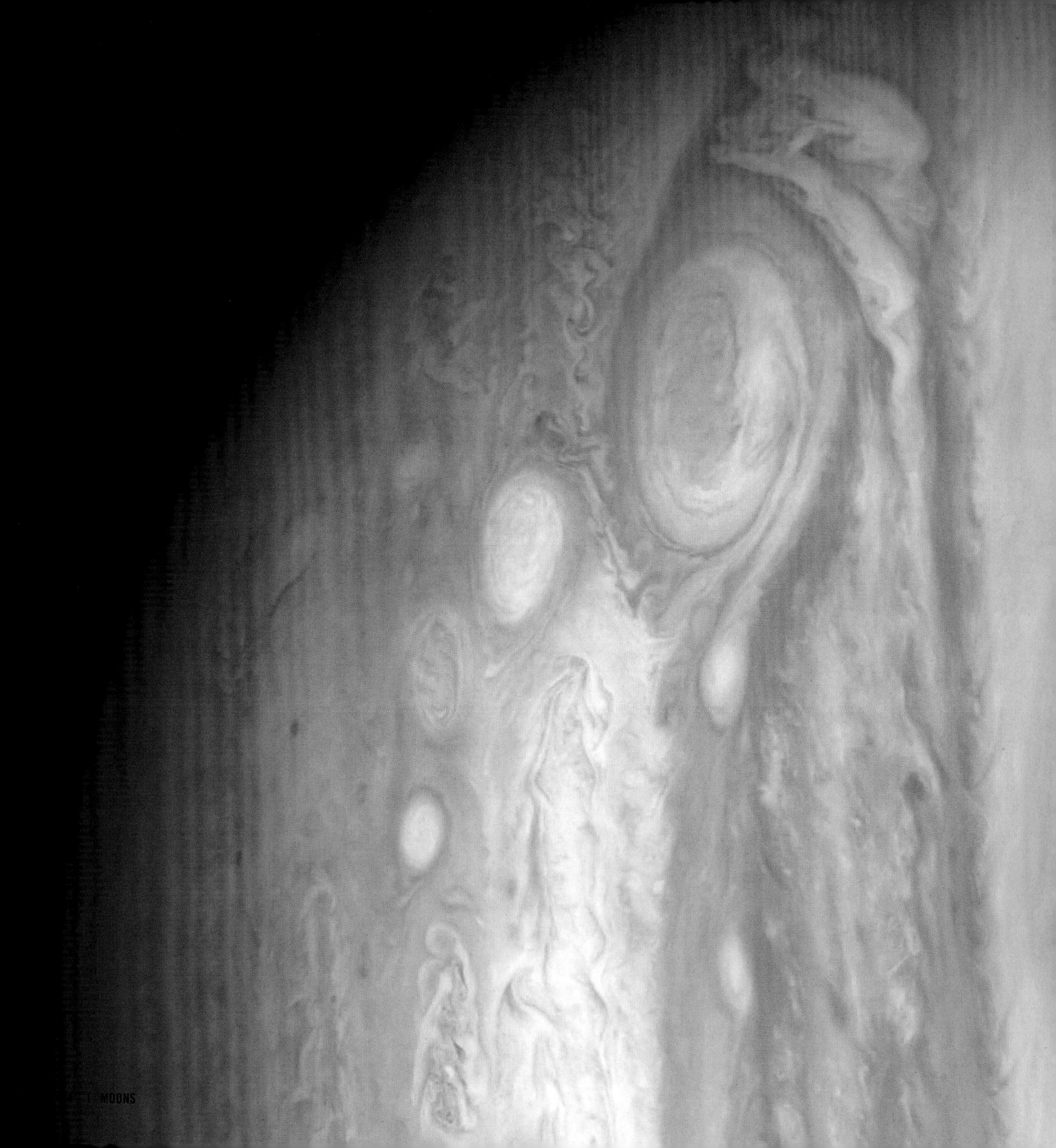

IO IN FRONT OF JUPITER. IMAGE TAKEN BY THE VOYAGER 1 SPACECRAFT. CREDIT: NASA/JPL/IAN REGAN.

EUROPA, GANYMEDE & CALLISTO

JUPITER'S OCEAN MOONS

After Io, the next moon out from Jupiter is Europa.

In appearance, Io and Europa couldn't be more different. While Io has a kind of hideous quality with its yellow surface and brown and red pockmarks, Europa is serene and beautiful, white with streaks of blue and red crisscrossing its surface. They're very different in composition, too: Io's hot, lava-coated rocks contrast sharply with Europa's frozen, icy surface.

When it comes to character, however, the two moons are much more alike. Like Io, Europa orbits close to mighty Jupiter and has sizeable moons orbiting beyond it. And like Io, Europa gets squeezed by fluctuating gravitational pulls that warm up its interior. The result is the same: the subsurface layer melts. Except on Europa, that molten layer is water instead of magma.

Io, as we know, is a rare rocky moon. Europa, on the other hand, is a much more common icy moon. It does have a rocky centre, but its surface is made almost entirely of water ice. At the extremely low temperatures you get out at Jupiter's distance from the Sun, water freezes as hard as granite and forms hard landscape – ice mountains and all. So when Europa gets squeezed by getting closer to and farther from Jupiter, the friction created inside the moon heats up the subsurface ice and turns it into liquid water.

Planetary scientists had theorised that this process might be happening on both rocky Io and icy Europa before it was ever observed. It's basic physics after all – gravity, friction and melting are all pretty simple concepts. But theories are just theories until there is evidence, and that evidence first came from the Voyager probes.

NASA's Voyager mission sent two spacecraft through the Jupiter system in 1979, including fly-bys of several of its moons. The images of Io that Voyager 1 and 2 radioed home showed active volcanic eruptions across its surface. This was groundbreaking, as it was the first time active volcanism had been seen on another world.

It was also concrete proof that there was indeed a layer of magma under Io's outer shell. From Europa, the Voyager probes returned images of a smooth surface – in fact, the smoothest surface in the entire solar system. Europa doesn't have big mountains, deep craters or canyons; there is just a network of stripe-like cracks in its icy shell. This smoothness suggested that Europa's surface was being regenerated in some way; most likely, by the presence of a subsurface liquid layer.

When a planet or moon has a hard exterior floating on a warmer liquid layer, the floating exterior moves around and crashes into itself at different points. On Earth and elsewhere, this movement creates mountains and other land formations. This is one of the main reasons why Earth – or Europa – isn't covered in craters from ancient impacts.

The liquid layer also sometimes bursts through the hard exterior, bringing fresh new material onto the surface. This process is called volcanism when it involves rock and magma, and cryovolcanism when it involves ice and water in cold places like icy moons.

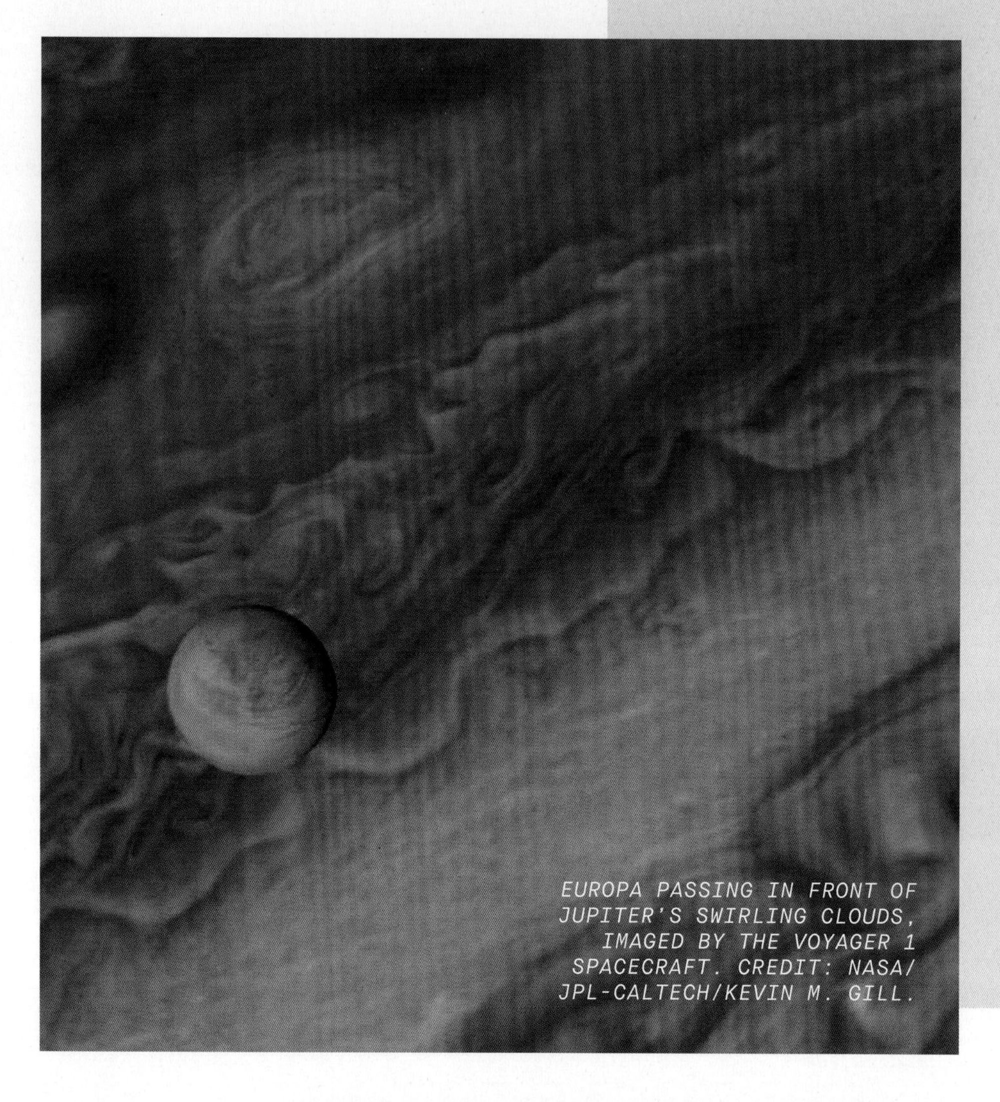

EUROPA PASSING IN FRONT OF JUPITER'S SWIRLING CLOUDS, IMAGED BY THE VOYAGER 1 SPACECRAFT. CREDIT: NASA/ JPL-CALTECH/KEVIN M. GILL.

NASA's Galileo mission returned to the Jupiter system in the 1990s and early 2000s, equipped with specific instruments for examining the moons as well as the planet. When it turned these instruments towards Europa, Galileo found that it was disrupting Jupiter's magnetic field in a way that suggested there was some kind of electrically conductive fluid under its surface – and as we know from horror stories about swimming during a lightning storm, water conducts electricity very well. This was another hint that Europa might have liquid water underneath its icy shell.

Although scientists can't be completely sure until there is direct evidence, they are fairly confident that Europa has subsurface oceans – big ones. The data suggest that under about 25 kilometres of ice, Europa may be holding an ocean 60–150 kilometres deep. Remember, Earth's oceans only stretch down about 11 kilometres at their deepest. Even though Europa is smaller than Earth's Moon, its oceans are thought to contain twice as much liquid water as all of Earth's oceans and lakes combined.

The implication of this is significant. On Earth, every single life form we've found has one thing in common: a dependence on liquid water. Life as we know it can survive all kinds of extreme conditions, especially once you get down to the level of microbial life. There are bacteria that thrive in pools of acid, around boiling hot hydrothermal

SIDEBAR

WHAT IS LIFE?

The only life we know anything about is the life we have here on Earth, so how can we extrapolate to define life more broadly when looking for it elsewhere in the cosmos?

Biologists have identified things that life forms on Earth have in common - things like having ordered structure, reacting to stimuli, metabolising nutrients into energy, growing, reproducing, and others - but this list doesn't perfectly unite all living things or exclude everything we consider inanimate. Fire transforms energy, for example, and crystals have ordered structure and can grow. Some living things grow too slowly to measure, or go through periods of dormancy where they stop metabolising or reacting to stimuli altogether. So even knowing what criteria have to be met for a thing to be considered living is complicated.

Part of the search for life beyond Earth involves looking for the chemical compounds that serve as the building blocks of life as we know it: carbon compounds, DNA and RNA, amino acids, etc. But again, it's not that simple - extraterrestrial life might be based on totally different chemical building blocks. Even water, which sustains all life on Earth, might not be the basis of alien life. Ammonia, methane and ethane have all been proposed as potential substitutes for water as a life-sustaining solvent.

The possibilities of what life might look like beyond Earth are infinite, and we don't know what we'll find when we do eventually get some confirmation that aliens are out there. But for now, the most logical thing to do is to focus our search on the kind of life we know how to recognise - hence our obsession with watery planets and moons.

vents at the bottom of the ocean, inside nuclear reactors – you name it. But every single living organism on Earth needs liquid water. From animals to plants to viruses, every living thing on Earth uses water to dissolve substances and carry them into and out of cells. This is how we turn food into energy and get rid of waste in our systems, among other life-sustaining functions. Because water is pretty much the only thing all life forms on Earth have in common, we focus on places where liquid water is present when looking for life beyond our planet.

For a long time, it looked like liquid water might not exist anywhere in the solar system beyond Earth. The Moon has water ice, but without an atmosphere keeping in heat, it's too cold for that ice to ever melt. Mars used to be wet, but has long since dried up. Even so, Mars has been the focus of the search for life beyond Earth for decades because it's been the closest thing to a watery world we've been able to find – plus, it's nearby. At one point, scientists thought they'd found signs of briny liquid water in the Martian soil, but the evidence for it was far from concrete and Mars is still considered a desiccated world.

Overall, no other planet or moon shows any promising signs of having liquid water right on its surface. So when Europa showed us that water could be hiding beneath the surface, this cracked the search for life in the solar system wide open. As you'll soon find out, Europa is not the only world with a secret subsurface ocean. Jupiter's moons Callisto and Ganymede, Saturn's Titan, and even distant Neptune's moon Triton, all show signs of possible subsurface oceans. If the idea that all life needs water applies beyond Earth, the number of places to go looking for it is a lot higher than we originally thought.

Europa's subsurface oceans may also have a few other life-friendly qualities. For one, they offer protection from radiation. Jupiter churns out an enormous amount of radiation, making the exterior of any of its moons unlikely to foster life. If you stood on Europa's surface, for example, you'd be exposed to a fatal amount of radiation within a day. But having 30 kilometres of ice over your head would definitely be enough to protect you. Under that icy shell, life as we know it would be able to flourish.

Europa's oceans also appear to be salty, which bodes well for possible life.

One piece of evidence for salty Europan oceans is the colour of the stripes that crisscross the moon's surface.

Scientists think that these stripes – which cover the moon and stretch as wide as 20 kilometres and as long as 1600 kilometres – may be reddish in colour because of salt water. The idea is that as Europa's crust moves around, it cracks, letting salty water seep through and reach the surface where it is exposed to Jupiter's intense radiation. Experiments have shown that salty water turns a reddish-brownish colour like this when it's exposed to the levels of radiation that are blasting Europa.

SIDEBAR

THE GOLDILOCKS ZONE

For a long time, astronomers had a cosy little term for the area around a star that was not too cold, not too hot, but just right for liquid water to exist: the Goldilocks zone. Or, if you want the more official and less fun name: the circumstellar habitable zone.

This concept was based on the observation that Earth was the only planet in our solar system with liquid water on its surface, suggesting that we were at just the right spot considering the size of our Sun. This led scientists to look for planets in Goldilocks zones around other stars as the prime places where we might find surface water and, therefore, life.

The discovery that Europa and other icy moons have subsurface liquid water has since rendered the Goldilocks zone a lot less special. When looking for life in the universe, we now have many more options to consider.

THE SCARRED SURFACE OF EUROPA, A COMPOSITE OF IMAGES FROM THE GALILEO SPACECRAFT. CREDIT: NASA/JPL/UNIVERSITY OF ARIZONA.

If this little moon's waters are indeed salty, that might mean that they are in direct contact with a rocky sea floor below, which might have hydrothermal vents (cracks in the sea floor). This is how Earth's oceans get most of their salt. Water seeps into these cracks and gets heated up by magma from the Earth's core, then comes blasting back out into the ocean. This brings heat into the rest of the ocean and causes a bunch of chemical reactions that release salt into the water, among other things.

Hydrothermal vents are also a big deal on Earth because they're among the top explanations of how life originated on this planet. The basic ingredients for making life seem to be water, organic compounds and an energy source. The heat coming out of hydrothermal vents could have provided the energy needed to turn non-living ingredients into life. And indeed, the oldest physical evidence of life ever found was fossilised microorganisms that lived in hydrothermal vents about four billion years ago, soon after the oceans themselves had first formed. Life still flourishes around hydrothermal vents today.

If Europa also has hydrothermal vents in its oceans, it might have the same conditions that allowed life to develop on Earth. So if we're looking for life, Europa is well worth exploring.

The issue is that 30 kilometres of rock-hard ice would be extremely difficult to drill through. Even a ragtag crew of oil drillers led by Bruce Willis might not be equipped for such a job. Planetary scientists have proposed missions that would try to bore through all that crust to send a probe down into the waters below, but none of them have yet been viable.

In 2012, the Hubble Space Telescope made a discovery that changed the game: it spotted what looked like a 200-kilometre plume of (probably) water vapour spewing out of Europa's south pole.

If indeed there are plumes of ocean water shooting through Europa's icy crust, we could potentially send a spacecraft to fly through them, collect samples of the plumes and analyse them to look for signs of life. This would be a lot easier than drilling through 30 kilometres of ice. But it's possible that these plumes aren't coming from the warm, salty, radiation-free subsurface ocean; they could just be coming from pockets of water within the crust, which lack all the life-friendly conditions of the oceans.

As is so often the case with space exploration, the little bit we've learned about Europa has opened up many more questions that can only be answered by sending another spacecraft to study it up close. Luckily, there are plans in place to do just that. NASA is aiming to send its Europa Clipper mission there in 2030, which includes a specific mission objective to fly through Europa's plumes. The European Space Agency is also working on its Jupiter Icy Moons Explorer mission (aka Juice), which would study Europa along with the other icy moons Ganymede and Callisto, starting in 2031. Both of these missions would be able to spend years studying Europa, potentially answering a lot of our most pressing questions about its tantalising oceans.

THE RUST-COLOURED STRIPES THAT COVER EUROPA'S SURFACE. IMAGE TAKEN BY THE GALILEO SPACECRAFT. CREDIT: NASA/JPL-CALTECH/KEVIN M. GILL.

JUPITER'S MOON EUROPA, IMAGED BY THE JUNO SPACECRAFT. CREDIT: NASA/JPL-CALTECH/SWRI/MSSS/KEVIN M. GILL.

AN IMAGINED VIEW OF AN EXPLORER IN THE WATERS BELOW EUROPA'S CRUST. CREDIT: GORDON AULD.

Continuing our trajectory out from Jupiter, the next major moon we come to is Ganymede.

Ganymede is the largest moon in the solar system, even bigger than the planet Mercury. It's also the only moon in the solar system that has its own magnetic field, complete with beautiful auroras like the ones we see in Earth's skies in the far north and far south.

Like Europa, Ganymede is made of rock with a thick outer layer of ice. But unlike its icy sister, Ganymede doesn't show any signs of a freshly resurfaced surface. When Voyager 1 and 2 passed Ganymede in 1979, they saw that it was covered in craters; it had experienced billions of years of impacts without anything to wipe away the evidence. This meant that it likely didn't have a layer of liquid underneath its icy shell; however, planetary scientists still suspected that interior-warming gravitational squeezing should be happening to some extent on Ganymede for the same reason it happens on the other Jovian moons (another term for moons of Jupiter).

NASA's Galileo mission got the chance to investigate while it was in orbit around Jupiter, flying past Ganymede several times and taking photos and collecting other data. Galileo's measurements of Ganymede's magnetic field suggested that there might be something going on underneath the surface – specifically, that there might be one or more layers of salty, liquid water under an extremely thick crust. Follow-up observations by the Hubble Space Telescope in 2015 backed up this theory.

From what we can tell, Ganymede might have several ocean layers, stacked up with layers of ice in between them. The highest level of ice is the crust, which might be about 150 kilometres thick. The lowest ocean level could be 800 kilometres below that, just above the first layer of the moon's rocky interior. Throughout this stack of oceans, this humongous moon likely has the most liquid water of any body in the solar system.

The thickness of Ganymede's crust would explain why it looks so different from Europa. The 150 kilometres of crust is too thick to break up and float around on the liquid layer, so you don't see any cracking or resurfacing. Still, Ganymede can't hide its water from the prying eyes of science!

Because Ganymede might have that magic combo of warmth, water, radiation protection and contact with the sea floor, it's on the list of places in the solar system where life could potentially exist. But because the outer crust is so heroically thick, there is no way to get through it to the liquid layers, and no helpful plumes are spewing out of it. That knocks Ganymede down the list of places to visit, but the European Space Agency's Juice mission will still learn a lot about this mega-moon when it visits the Jupiter system in the 2030s.

GANYMEDE IN COLOUR, IMAGED BY THE VOYAGER 1 SPACECRAFT. CREDIT: NASA/JPL/TED STRYK.

The outermost of Jupiter's major moons is Callisto. It's the third-largest moon in the solar system, only about 60 kilometres smaller in diameter than Mercury.

Measurements of Callisto's density show that it's about half rock, half water ice. But unlike Europa and Ganymede, Callisto's rock and ice never fully separated into distinct layers. Instead, it seems to be a sort of rock-ice blend all the way through, with the amount of rock increasing as you get deeper.

Because Callisto is a bit farther from Jupiter and doesn't have any big moons beyond it, it doesn't experience the kind of gravitational fluctuations that warm up the interiors of the other Galilean moons. As a result, Callisto doesn't have any tectonic activity or volcanism – no internal processes to alter its surface in any way. In fact, Callisto is thought to have the oldest surface in the solar system.

But even though nothing internal is reshaping it, Callisto's surface has definitely been influenced by outside forces over the moon's 4.5-billion-year existence. It has more impact craters than any other body in the solar system and any new impactor would almost be guaranteed to hit an existing crater. This is in part because none of the craters have ever been erased. Geologic activity (the main crater-eraser) is absent on Callisto, so every crater that gets made stays there for eternity. Callisto is also particularly prone to impacts because of its location. Jupiter's gravity pulls in wandering asteroids, comets and meteoroids that can result in collisions with the moons.

Despite showing no signs of geological life, there are some signs that Callisto has liquid water deep below its surface. Once again, NASA's Galileo spacecraft was responsible for finding signs of hidden water while it was studying the Jupiter system.

Unfortunately, the possibility of water under Callisto's crust doesn't necessarily mean it could be teeming with life. What Callisto is missing is an energy source within that ocean. Although scientists are still debating it, there is no decisive evidence to prove that Callisto's ocean waters interact with a rocky sea floor, or that there are hydrothermal vents providing heat to those waters. So although Callisto is still considered an ocean moon, it is not a hot destination in the search for life.

JUPITER'S MOON CALLISTO, IMAGED BY THE GALILEO SPACECRAFT. CREDIT: NASA/JPL/DLR.

ENCELADUS

THE HOSPITABLE MOON

For a long time, all we knew about Saturn's moon Enceladus was that it was there, and that it was very bright.

It had been discovered in 1789, but could only be seen as a little dot through a telescope until the Voyager spacecraft flew through the Saturn system in the early 1980s. These fly-bys yielded the first close-up images of Enceladus, but these didn't spark a lot of exciting discoveries beyond awarding Enceladus the title of brightest object in the solar system. All we saw was a blindingly white snowball world.

A couple of decades later, NASA's Cassini spacecraft became the first mission to enter into orbit around Saturn to study the giant planet and its rings and moons up close over many years. When Cassini flew by Enceladus in 2005, the little icy moon finally got its chance to impress us.

At the south pole of Enceladus, Cassini observed some very strange features. The whole region is covered in a series of huge, long gashes that the mission team dubbed 'tiger stripes'.

These stripes are enormous fractures in the surface ice, with long mountainous ridges along either side. Around the stripes the ice appears young, far less cratered than the ice on the moon's northern hemisphere. The explanation for this area's rejuvenation came pretty quickly.

As Cassini continued to observe Enceladus, it saw enormous plumes shooting out of the south pole, beyond anything seen on Jupiter's icy moons. These were extremely large, active cryovolcanoes blasting huge amounts of water, ice, organic molecules and other material into space. Cassini counted more than 100 of these cryovolcanoes in the south pole area alone.

These volcanoes collectively shoot about 200 kilograms of material (mostly water vapour) per second. All of this material exits the volcanoes extremely fast, moving at about 400 metres per second and reaching hundreds of kilometres high.

While this amount of cryovolcanism is a lot to wrap your head around, it makes a lot of sense when you consider two factors: Enceladus' size and its location.

Enceladus is only about 500 kilometres across. Europa, for comparison, is over 3000 kilometres across, and the rest of Jupiter's major moons are even larger. Scientists didn't expect Enceladus to have liquid water in it, since bodies this size and this far from the Sun are normally frozen all the way through.

But, like Jupiter's geologically active moons, Enceladus orbits a humongous planet and has other big moons on either side of it. You know the story from here: the other moons tug on Enceladus, making its orbit around Saturn a bit uneven; it gets squeezed gravitationally and its interior heats up. The effect on Enceladus, though, is very different because it's significantly smaller than its Jovian cousins – and small size means low gravity.

When cracks form in Enceladus' icy shell and the liquid water beneath gets pushed out, there's a lot less holding it back. This is the same reason that volcanoes on a little moon like Io shoot plumes so dramatically high into space compared to the volcanoes that we're used to here on Earth. With less gravity to pull it back down, the force of an eruption can really blast stuff – be it lava or water – in huge quantities, at breakneck speeds, and to astonishing heights.

A lot of the material erupting from Enceladus' south pole falls back down as snow, giving the moon its bright, white coating.

But the lighter particles escape the moon's weak gravity and get swept up into orbit around Saturn. It turns out that Saturn's widest and outermost ring (called the E ring) is mostly made of material that was spewed out of Enceladus' interior.

For obvious reasons, the discovery of massive plumes of water shooting from Enceladus was big news. Over the next few years, NASA had Cassini fly past Enceladus a few more times to find out more about this suddenly exciting and important little moon.

When NASA built the Cassini spacecraft, they didn't know that Enceladus had a subsurface ocean spewing out of its south pole, so there were no specific instruments on board to collect and analyse that water. But when the plumes were discovered, the mission team knew they had to get a sample. Luckily, space scientists and engineers are a scrappy bunch of problem-solvers and are famous for coming up with ingenious ways of doing things on the fly. Since Cassini was intended to orbit near Saturn and its rings and moons but not touch them, it was mostly equipped with tools for measuring things from a great distance. Some of these long-distance tools were put to use to analyse the plumes of Enceladus, but scientists were also able to adapt the function of an instrument that was designed to collect and analyse tiny grains of cosmic dust.

For its first encounter, Cassini skirted the edges of the plumes to see if they would damage the spacecraft. When it came out unscathed, the spacecraft made another pass through a denser part of the plumes. This was where a lot of the exciting data showing the chemical composition of the plumes were found.

Cassini's dip through the plumes showed that Enceladus might just about have the best life-friendly conditions you could ask for. Beneath about 30–40 kilometres of radiation-shielding ice, there is a deep, salty ocean that comes into direct contact with a rocky seabed. That seabed almost definitely has active hydrothermal vents which bring energy into the ocean, helping to create complex molecules like amino acids that are the building blocks of life.

What's almost painfully tantalising about this discovery is that it's possible that life may already exist in Enceladus' oceans and may even have been present in the plumes that Cassini sampled. Unfortunately, no amount of creative problem-solving could equip the spacecraft's instruments to detect this. Figuring out whether something is living is a whole different process from measuring its mass or chemical composition. So until we send another mission to Enceladus to specifically look for signs of life, we won't know for certain whether this is just a habitable world, or an actual inhabited world.

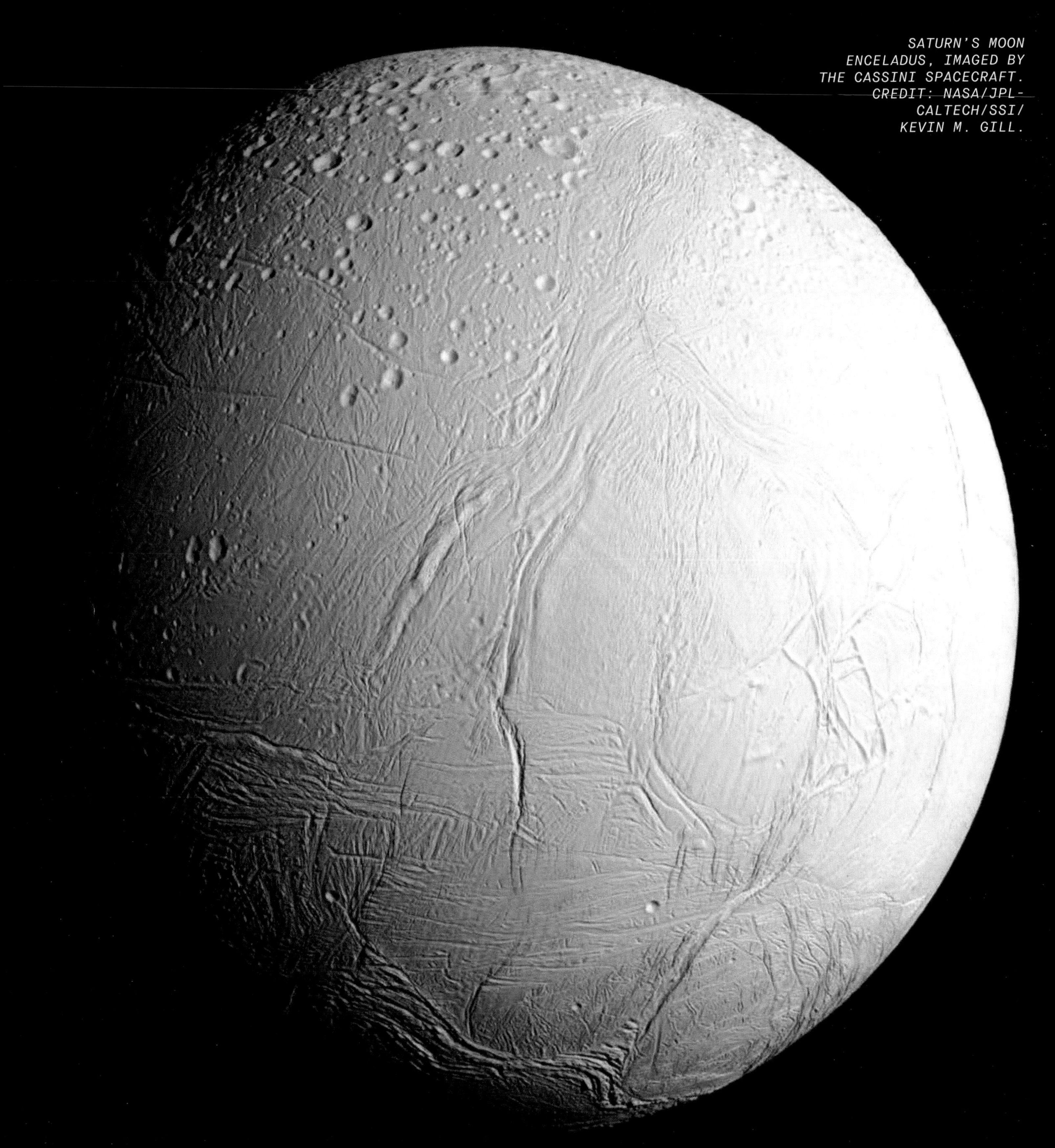

SATURN'S MOON ENCELADUS, IMAGED BY THE CASSINI SPACECRAFT. CREDIT: NASA/JPL-CALTECH/SSI/ KEVIN M. GILL.

NEAR AND FAR VIEWS OF PLUMES OF WATER ERUPTING FROM THE SOUTH POLE OF ENCELADUS. IMAGES CAPTURED BY THE CASSINI SPACECRAFT. CREDIT: NASA/JPL-CALTECH/SSI.

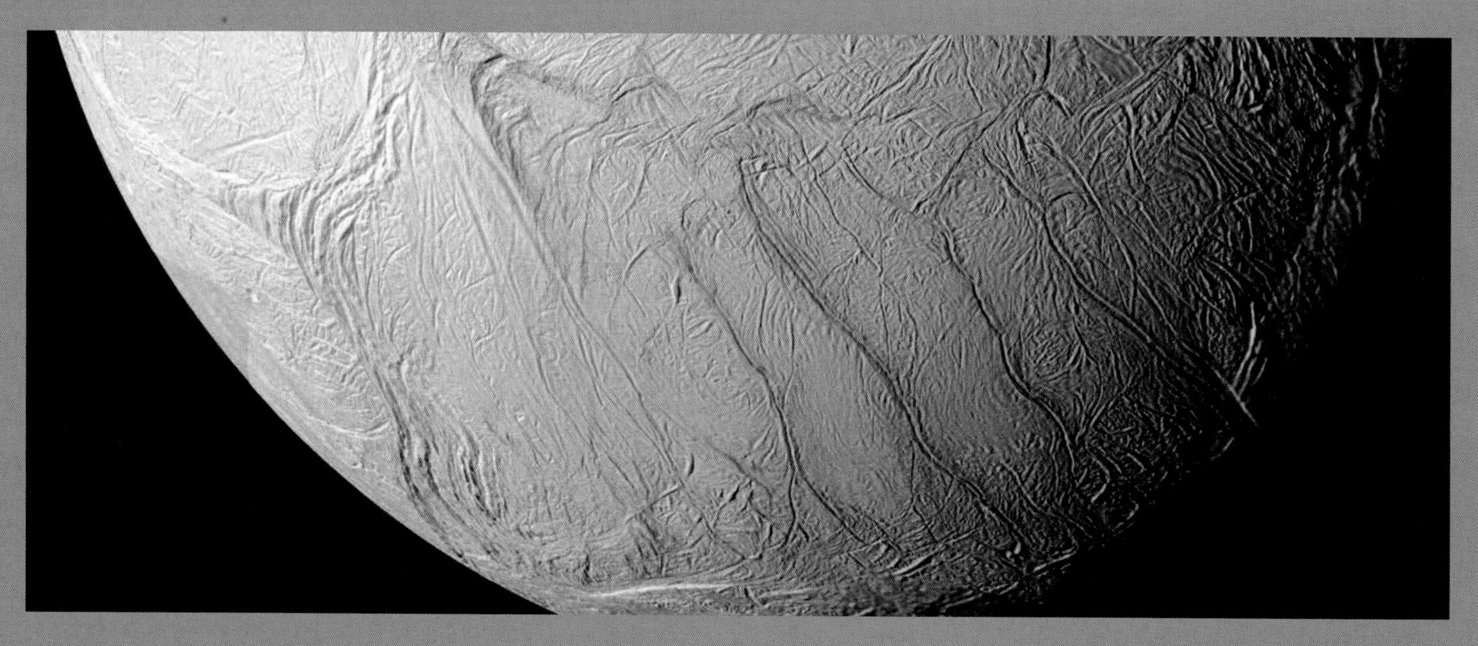

THE TIGER STRIPES' ON ENCELADUS' SOUTH POLE, IMAGED BY THE CASSINI SPACECRAFT. CREDIT: NASA/ESA/JPL-CALTECH/SSI.

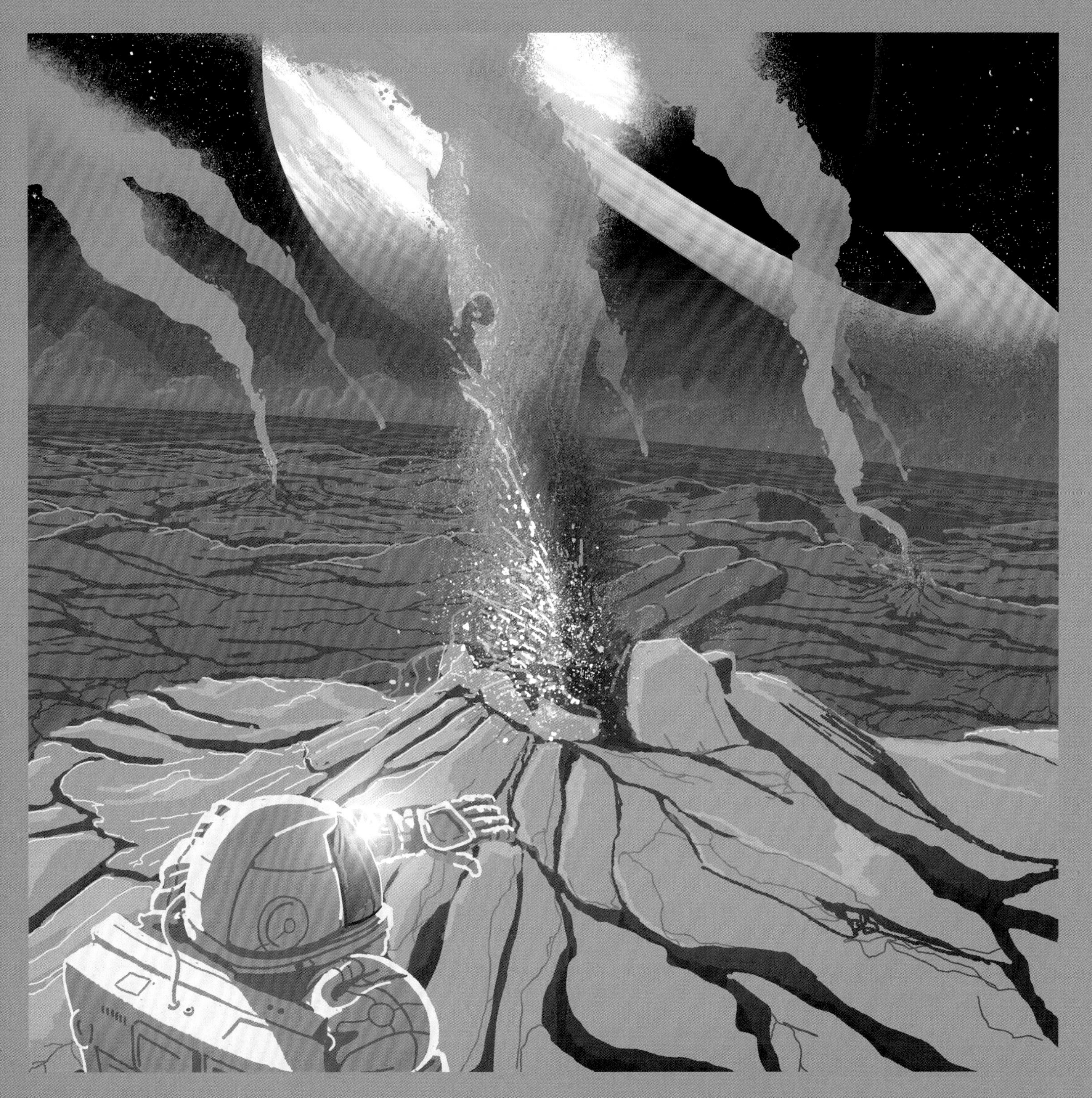

AN IMAGINED VIEW OF AN EXPLORER ENCOUNTERING AN ERUPTING PLUME ON ENCELADUS. CREDIT: GORDON AULD.

THE *CAPTURED* PLANET

TRITON

About three billion kilometres from Enceladus, we find Neptune's moon Triton.

Triton is named after a mythological merman (who you may remember as Ariel's dad in The Little Mermaid*), in line with the pattern of naming the moons of Neptune (the Greek god of the sea) after minor water deities in Greek mythology.*

Triton is the largest of Neptune's 14 known moons. It's the seventh-largest moon in the solar system, and it is more massive than all the other moons smaller than itself combined. It's also super bright, reflecting 60–95 per cent of the sunlight that reaches it (compared to Earth's Moon, which only reflects 11 per cent). That being said, Triton is so far away from the Sun that it gets 900 times less sunlight than Earth.

Parts of its surface also spend decades at a time in darkness. Because it's on an extreme tilt relative to the Sun, Triton's north and south poles take turns pointing almost directly at the Sun and then being in total darkness. These periods aren't brief: Neptune takes 165 years to orbit around the Sun, so each season on Triton lasts for nearly 41 years.

Triton's main claim to fame is that it's likely a dwarf planet that was captured by Neptune's gravity. The main reason scientists think Triton started off as its own planet is that Triton orbits Neptune the wrong way. If you're viewing north as the top (as we tend to do), Neptune spins counterclockwise on its axis. And because planets and their moons generally all form from the same swirling mass of stuff, the direction of swirl should remain the same – that is, the moons should orbit the planet in the same direction the planet spins. Because Triton orbits Neptune in a clockwise direction, it's unlikely that it formed alongside its host planet.

What's more likely is that Triton formed on its own somewhere in the Kuiper Belt, a huge ring of smaller, icy objects (mostly comets) that orbit the Sun beyond Neptune. Pluto, the most famous of the dwarf planets, is out in the Kuiper Belt, and actually has a lot in common with Triton: similar mass, diameter and surface materials. Objects in relatively dense areas like the main asteroid belt and the Kuiper Belt gravitationally pull on each other as they pass, affecting each other's trajectories through space. (This is why an asteroid sometimes leaves the asteroid belt and comes careening towards Earth to wipe out the dinosaurs.)

Out in the Kuiper Belt, Triton likely got nudged by another large object into a trajectory that brought it close enough to Neptune to get captured by its gravity and pulled into its orbit. Upon its arrival, Triton probably wreaked havoc on Neptune's pre-existing moons. Neptune has fewer moons today than you'd expect for a planet its size, and scientists think it's likely that when Triton entered the Neptune system it smashed into some of its moons and destroyed them, disrupting other moons' orbits with its gravity and causing them to smash into and destroy each other.

Triton certainly made a grand entrance, and will probably have a grand demise as well. Neptune's gravity is slowly pulling it inward, and in a few million years it'll get so close that Neptune's gravity will actually tear it apart. This sounds like a sad fate, but there is an upside. Neptune already has a tiny, faint set of rings, but when Triton breaks up, its material will likely spread out into a set of rings more like Saturn's.

Despite it being such an intriguing world, we've barely explored Triton. Only one spacecraft has visited the Neptune system: the Voyager 2 mission. This visit happened on 25 August 1989 after Voyager 2 had already flown past Jupiter, Saturn and Uranus.

UP AND DOWN IN SPACE

The whole concept of up and down is completely relative. We Earthlings consider the north 'up' and the south 'down', almost entirely for shady colonial reasons – the world's most powerful nations have historically, at least in recent centuries, been located in the Northern Hemisphere, so they deemed it the top of the world.

This is why every image of planets, moons and anything else in space shows the north pole at the top and the south pole at the bottom. But this isn't how space is oriented. Up and down are relative concepts. Some other alien species that let colonists from its southern region run amok probably shows all of its space pictures the opposite way from us.

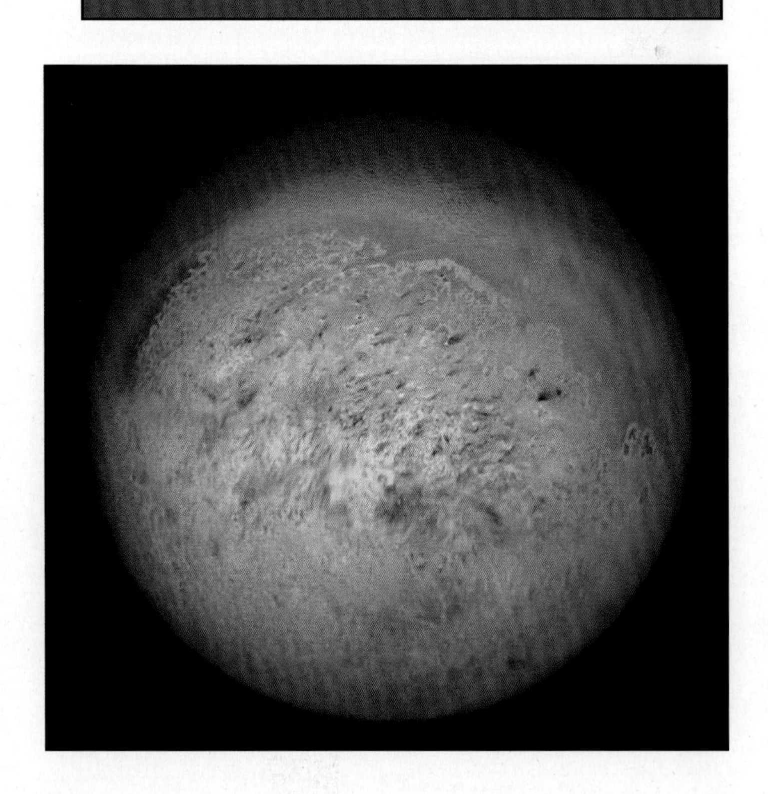

ANOTHER VIEW OF TRITON CAPTURED BY THE VOYAGER 2 SPACECRAFT. CREDIT: NASA/JPL-CALTECH.

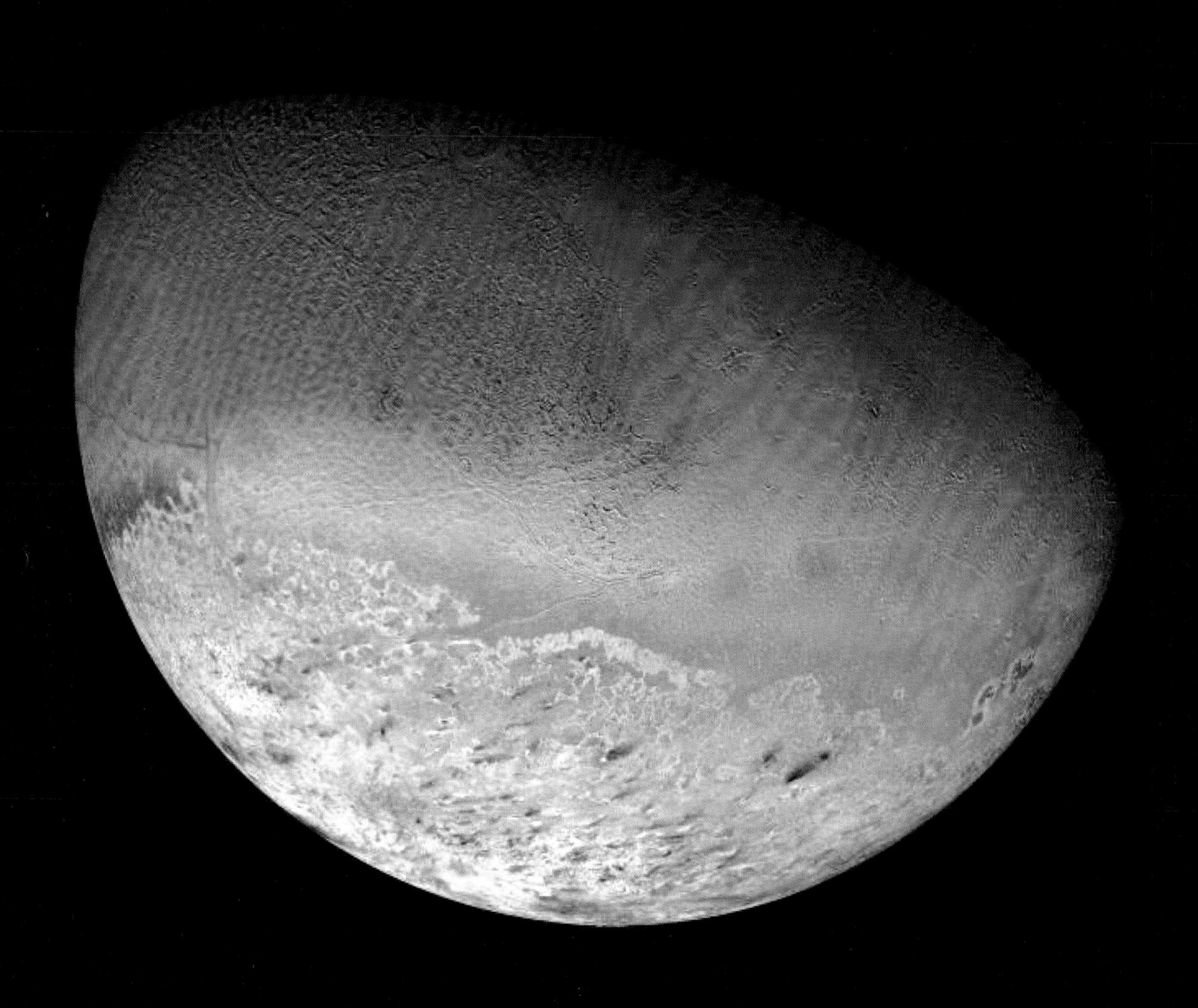

NEPTUNE'S MOON TRITON,
IMAGED BY THE VOYAGER 2 SPACECRAFT.
CREDIT: NASA/JPL/USGS

SIDEBAR

THE VOYAGER MISSION

The Voyager mission is among one of the coolest things we've ever done in space. The mission was designed to take advantage of a very rare alignment of the outer planets in the late 1970s and 1980s – with Jupiter, Saturn, Uranus and Neptune all at similar places in their orbits around the Sun – that happens every 175 years or so. NASA decided to make the most of this alignment by sending two identical spacecraft on this multi-planet journey, taking slightly different routes so they could make different observations. Voyager 1 and 2 were launched in 1977 just two weeks apart.

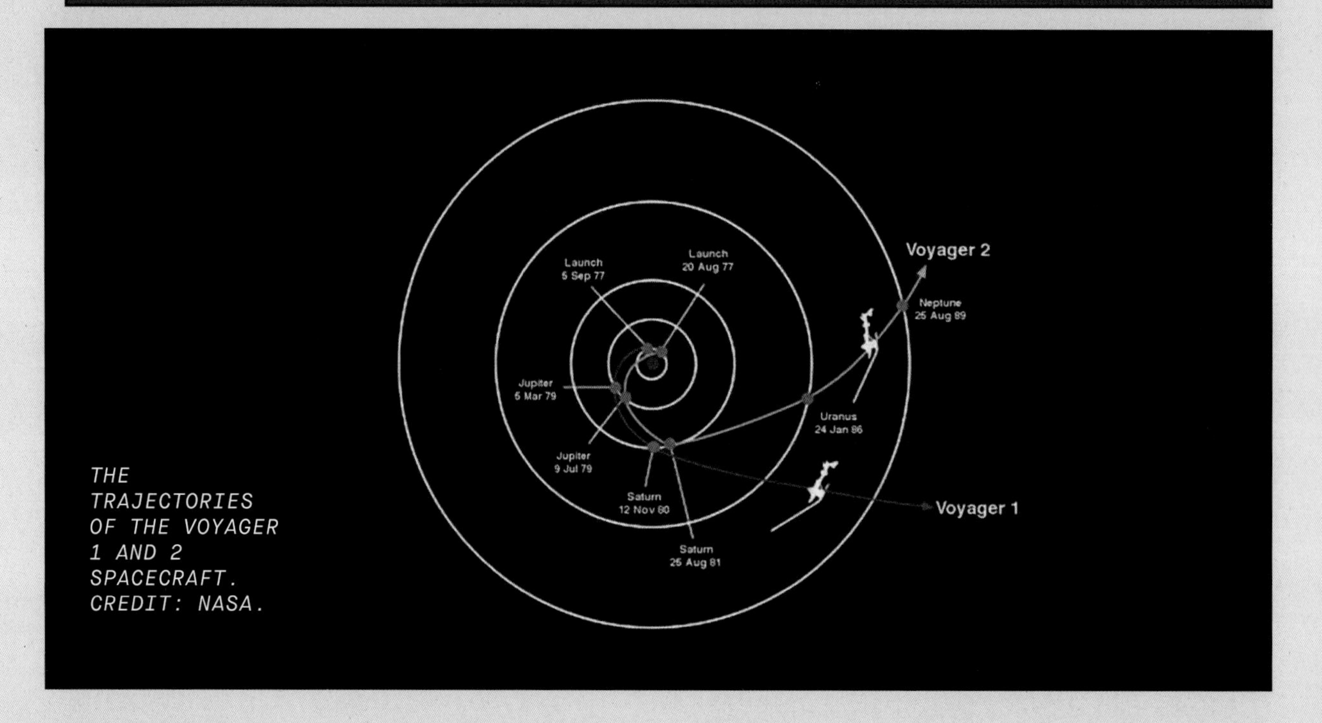

THE TRAJECTORIES OF THE VOYAGER 1 AND 2 SPACECRAFT. CREDIT: NASA.

The twin spacecraft were originally just going to visit Jupiter and Saturn due to tightening science budgets at the time. But after finishing its fly-by of Saturn in 1981, Voyager 2 had enough gas in the tank for a few more trajectory alterations, and because the Voyagers had been so successful thus far, NASA decided to extend Voyager 2's mission, sending it on to Uranus and then Neptune.

One of the remarkable things about the Voyager trajectories is that each planetary fly-by sped the spacecraft up. Every time the Voyagers passed a planet, they would swing around it, drawn in by its gravity but not enough to be pulled into its orbit, instead accelerating and shooting off in a slightly different direction. Engineers call this a gravity assist or slingshot manoeuvre, and these gravity assists allowed Voyager 2 to reach Neptune 12 years after it launched, instead of the 30 years it would have taken to get there without any planetary help.

The two sister spacecraft opened our eyes to the outer solar system, showing us places we'd only ever seen as dots in a telescope. They sent back images and data of worlds we'd never even dreamed of, including many of the weird and wonderful moons described in this book.

After their planetary fly-bys, the Voyagers just kept on going and are now officially in interstellar space, beyond the Sun's influence. Miraculously, at the time of writing they are both still communicating with Earth.

What's even more incredible is that the Voyager spacecraft could continue hurtling away from the solar system for millions of years. Because space is so huge and empty, it's extremely unlikely that either spacecraft will smash into a planet, get sucked into a black hole or pass too close to a star and disintegrate.

And because they're transmitting radio signals, it's possible that they could someday be detected by some kind of distant alien life that, like us, is listening for radio signals from space. The scientists working on the Voyager program thought about this possibility and put a message on each spacecraft for those hypothetical aliens. Each Voyager carries a 12-inch gold-plated phonograph record (this was the 1970s, remember) with pictures and sounds of Earth encoded into it. On its cover each one has instructions for playing the record and for finding the location of our planet, written in mathematical language that scientists figured any intelligent species should be able to decipher.

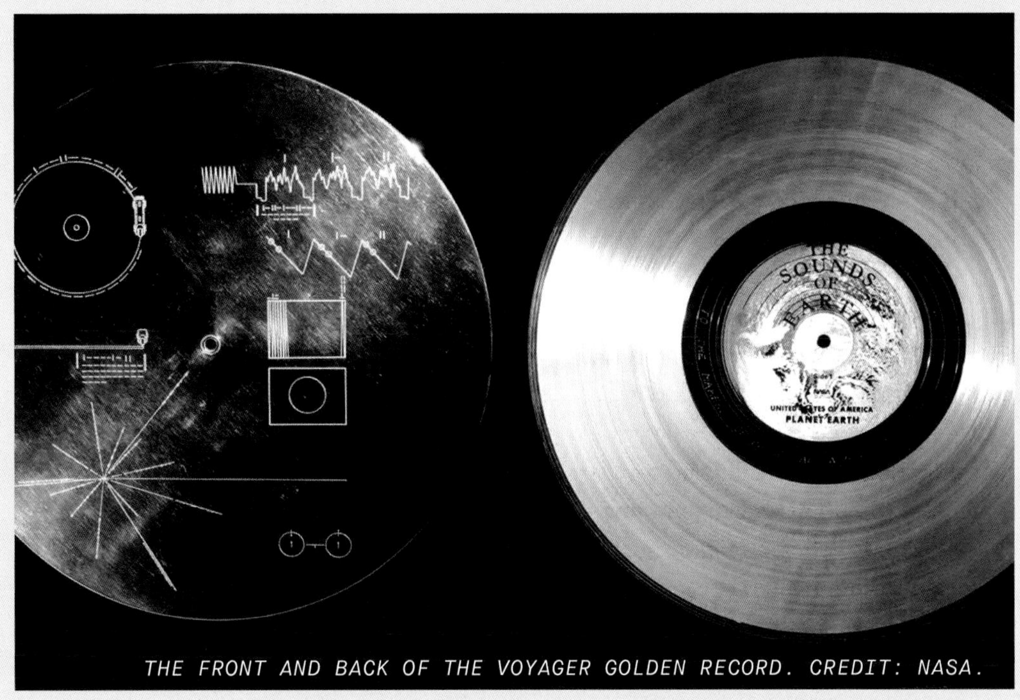

THE FRONT AND BACK OF THE VOYAGER GOLDEN RECORD. CREDIT: NASA.

Each contains images of animals, sounds of waves crashing, recordings of songs (mainly classical music and traditional music from around the world, with Chuck Berry's 'Johnny B. Goode' thrown in for good measure), photos of people from various cultures and all sorts of other things, to get a snapshot of what makes Earth, Earth.

The Voyager golden records are a time capsule of Earth, life and human culture.

So even when the Sun swallows up the Earth, which it will inevitably do, this little message in a bottle will continue to exist in the cosmos.

AN IMAGINED VIEW OF VOYAGER 2 PASSING TRITON AND NEPTUNE. CREDIT: GORDON AULD.

Voyager 2's encounter with Neptune took place over a matter of hours, since the spacecraft was moving at a breakneck speed of 90,000 kilometres per hour at that point. Its closest approach to Neptune was 4950 kilometres – which is pretty close in celestial terms (Earth's Moon is more than 380,000 kilometres from us). Voyager 2 took photos and measurements of Neptune and then sped on. Five hours later it flew by Triton, about 40,000 kilometres from the moon. The spacecraft took as many measurements as it could in the brief fly-by, taking pictures and studying about 40 per cent of Triton's surface, learning basically everything we now know about the moon. Then, just like that, Voyager 2 was gone again.

Despite making such a brief appearance, Voyager 2 managed to discover a whole lot about Triton. It took the first-ever images of the moon, and made some surprising observations – the biggest being that Triton is geologically active. Like most large moons beyond Jupiter, Triton's crust is made of water ice (although extra-cold Triton also has a surface layer of frozen nitrogen), and like many of the moons we've already discussed, Triton's crust seems to be dynamic, cracking and shifting and resurfacing itself. As usual, the most likely explanation is that a layer of liquid lurks beneath the hard crust.

The first hint of this subsurface ocean came from Voyager 2's observation that Triton doesn't have as many craters on its surface as you'd expect for a moon its size. If you hang out in space for long enough (like, billions of years), you eventually get smacked by a lot of wandering rocks. Scientists actually have a pretty good idea of the rate at which craters accumulate, and they can apply that rate to figure out the age of a planet or moon's surface based on how many craters it has.

This may seem like a rather slapdash way of ageing a planetary body, but it's held up under scrutiny. The Apollo missions brought back tons of rocks from the Moon in the 1970s that scientists have been able to age using radiometric dating in proper, Earth-based labs, and the numbers match up with the crater-counting results.

Because we know that all the stuff in our solar system formed at around the same time (about four billion years ago), if an object's surface is considerably younger then we can deduce that something has been refreshing it and churning up subsurface material.

Triton's dearth of craters isn't the only evidence for a subsurface layer. Volcanic activity of any kind is a huge clue that there is a liquid layer underneath the surface, and Voyager 2's observations of Triton found plenty of evidence of volcanism. As the spacecraft sped past, it saw more than 120 dark streaks on the surface of the moon's southern hemisphere, likely made as material spewed from volcanoes in the not-too-distant past. It also spotted the second-largest volcano in the solar system, which researchers named Leviathan after the biblical sea serpent.

Leviathan has a central volcanic crater about 100 kilometres in diameter, and the entire volcano is about 2000 kilometres wide at its base. It's also connected to two enormous cryolava lakes (which is essentially a really cool way of saying ice lakes).

The real excitement, however, came from evidence of present-day volcanic activity. In its brief fly-by, Voyager 2 spotted two cryovolcanoes that were actively erupting, including one that was shooting subsurface material eight kilometres up into the atmosphere.

Scientists are cautious about making big declarations before all the evidence is in, so even though Triton is made of water ice, researchers are careful to point out that the subsurface liquid might not be liquid water; it could be liquid nitrogen or something else altogether. But the possibility that Triton has a subsurface water ocean is plausible too. When Triton was first captured by Neptune's gravity, it would have created a lot of stress and friction within the moon, as it got pulled this way and that before settling into a stable orbit. All of that friction could theoretically have been enough to melt the water within Triton and keep it liquid for billions of years.

Whatever liquid it may be, the stuff that spews out of Triton's volcanoes has given the Neptunian moon another unusual characteristic: an atmosphere. Very few moons in the solar system have their own atmospheres, because it's generally difficult to keep one around. Even in the far reaches of the solar system, the Sun's radiation is powerful enough to strip the atmosphere off a moon or planet if it doesn't have a magnetic field protecting it. Magnetic fields are generated by liquid metal churning at the core of a planet or moon, and most moons are too small to sustain a liquid core. Ganymede is the only moon with its own magnetic field, and that's only because it's a true behemoth of a moon. Without a magnetic field to keep an atmosphere in place, the only way for a moon to have one is by continuously replenishing it. This is what Triton does – its cryovolcanoes bring gases as well as liquids up to the surface, and much of that gas lingers above the surface as an atmosphere.

Another noteworthy thing about Triton is that it has a surface unlike any other object in the solar system. Voyager 2's photos of its western hemisphere show a strange pattern that looks a bit like the skin of a cantaloupe – inspiring scientists to name it 'cantaloupe terrain'.

This melon-esque look is made up of polygon-shaped bumps in the icy surface, mostly about 10–30 kilometres across, with tall ridges in between them that can be up to several hundred metres high. If you suddenly found yourself on Triton, walking around on this terrain would be like making your way through some kind of hedge maze, but with cliffs of rock-hard ice instead of hedges.

Cantaloupe terrain doesn't appear anywhere else that we've seen, and because we've only observed it once during Voyager 2's brief fly-by, scientists still aren't sure what causes it. One possibility is that a layer of solid nitrogen ice lower in Triton's crust contracted in the past, making the outer layer above crack the way mud dries and cracks on Earth.

Altogether, we won't really understand Triton until we send another mission there to study it up close, and it'll probably be quite a while before that happens. At the time of writing, China is considering a mission to Neptune that would launch in 2030 and arrive by 2040, including a tiny spacecraft that would examine and ultimately smash into Triton to find out just what exactly is going on beneath the surface. Until then, Triton will keep its secrets to itself.

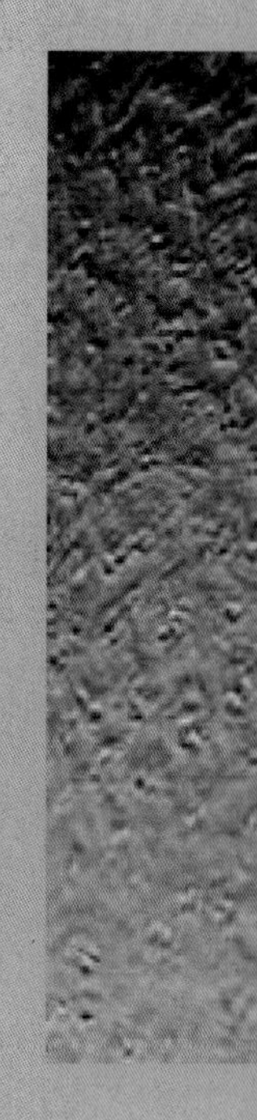

A COMPOSITE OF IMAGES OF TRITON CAPTURED BY THE VOYAGER 2 SPACECRAFT. CREDIT: NASA/JPL-CALTECH.

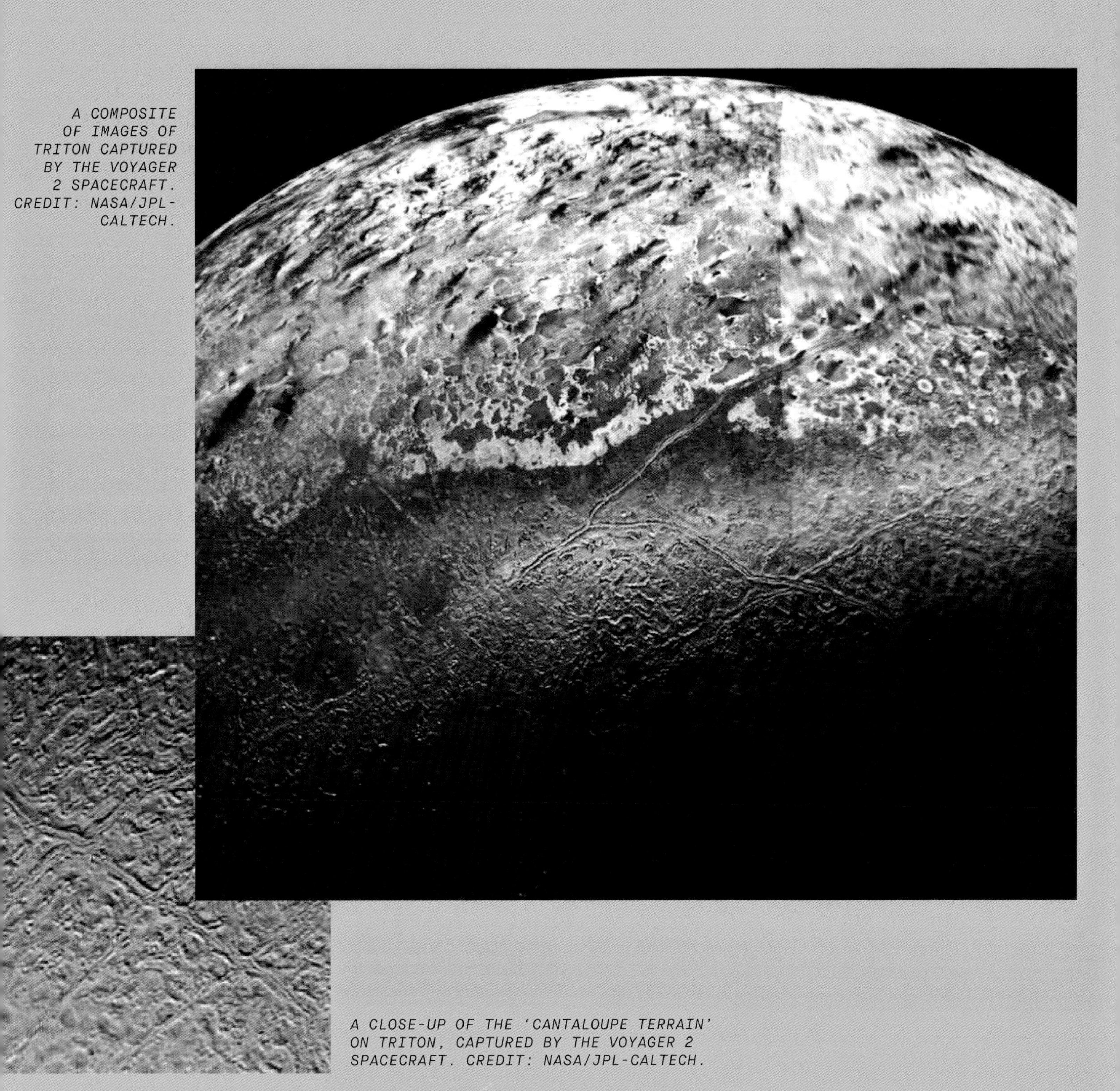

A CLOSE-UP OF THE 'CANTALOUPE TERRAIN' ON TRITON, CAPTURED BY THE VOYAGER 2 SPACECRAFT. CREDIT: NASA/JPL-CALTECH.

THE MOON

OUR VERY OWN MOON

Earth's Moon is our moon, aka THE Moon.

I'll admit that I used to think our Moon was unexciting compared to some of the others, but once you scratch below the surface (figuratively speaking), it's got a pretty fascinating story.

Scientists think that the Moon was formed in the early days of the solar system, roughly 4.5 billion years ago, when another young planet smashed into the Earth. This was the norm back then. Collisions are how everything in space comes to be, whether it's tiny particles of cosmic dust combining into slightly less tiny particles, or huge rocks crashing into each other to form even huger rocks. In the early solar system, planets formed through both big and small collisions.

This particular smash-up would have been very intense. The two colliding planets would have thrown huge chunks of super-heated rock out into space. Within a very short time – new research suggests it could have taken only hours, or as long as months – these chunks would have come together, attracted to each other's gravity, and formed the Moon.

SIDE BAR

LUNA

The Moon's Latin name is Luna. That's why we have the word 'lunar', which refers to things relating to the Moon. Unlike most other moons that are named after unrelated deities, the Moon itself has been considered a god figure by many cultures for millennia. The ancient Romans called this deity Luna, and like a lot of other moon deities she was a goddess, the female partner of the sun-god. Today, in many cultures, the Moon is still associated with femininity, mystery, emotion and the subconscious, among other deep and beautiful things.

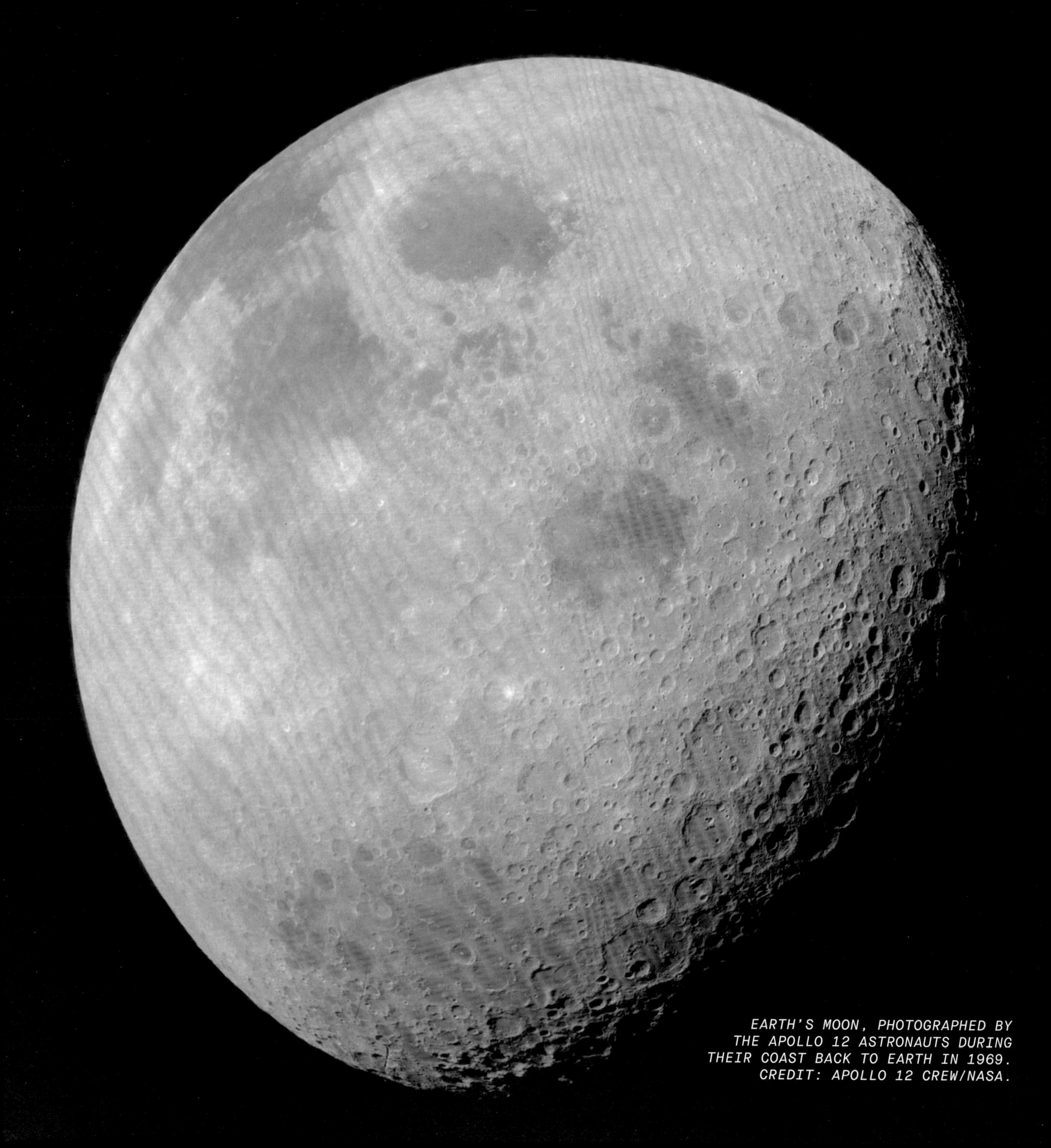

EARTH'S MOON, PHOTOGRAPHED BY THE APOLLO 12 ASTRONAUTS DURING THEIR COAST BACK TO EARTH IN 1969. CREDIT: APOLLO 12 CREW/NASA.

Now, I say 'would have' because it's not a certainty that this is how it all went down. Scientists still haven't concretely proven exactly how the Moon formed, but this is a leading theory. It also explains one strange feature of the Moon: the rocks in its outer layers all seem to have been molten at one point. And this is a remarkable thing to contemplate.

What we think happened is that the heat that came from the impact and the very rapid accretion (that is, the coming together of all the chunks) was so intense that the entire surface of the Moon became magma. And even though the formation of the Moon happened extremely quickly, its surface stayed molten for tens or even hundreds of millions of years.

For a period of time longer than anyone can really comprehend, the Moon was covered in an ocean of lava.

The Earth got pretty beat up by this impact too, of course. It didn't experience the same level of heat that the Moon got from its quick accretion, but the Earth's surface was a molten hellscape for a few millions years nevertheless.

Eventually, things cooled down. Minerals in the magma crystallised, turning solid. This happened in the lower depths first, with a crust forming on the surface later. In the middle, though, magma stuck around. Up until about 50 million years ago, including when the dinosaurs ruled the Earth, the Moon had volcanoes that spewed this magma out onto the surface from time to time. This volcanic activity would even have generated an atmosphere, probably about twice as thick as the current Martian atmosphere.

None of this lasted, and although there is still magma under the Moon's surface, it is too thick to bubble up through the crust. Still, when we look at the Moon today, we can still see evidence of its volcanic past. The familiar face of the Moon has dark splotches all over it, which scientists call plains, seas or maria. The lighter areas on the Moon are called highlands. The darker plains are made of basaltic rock, which erupted from volcanoes and settled into low-lying areas like impact craters.

Orbiting spacecraft have also found lava tubes, which are as cool as they sound. Back when volcanoes were spewing magma onto the Moon's surface, sometimes the top layer of the lava flow would cool down, with the remaining magma continuing to flow beneath in a tube-shaped passage. Eventually, the lava would all drain out, leaving the tube behind. Spacecraft can spot these from above when a portion of the tube collapses, leaving a hole that scientists aptly call a skylight.

We see lava tubes all over Earth and on Mars too, but the lunar ones are particularly enormous. The biggest lava tubes we've found on Earth, in the Kazumura cave on the island of Hawaii, are about the size of a typical tunnel you'd drive a car through. On the Moon, lava tubes can be 300–700 times that size. Some are tall enough to fit the world's tallest building inside them. Others are large enough to contain a whole city – and of course, there are lots of ideas out there about building human settlements inside lava tubes, since they'd offer protection from meteorites and solar radiation.

It's worth noting that almost all of the ancient volcanoes on the Moon are on the side that faces the Earth. One of Luna's neat tricks is that as it orbits the Earth, it rotates at exactly the right speed to always show the same face to the Earth. This is called tidal locking and it happens more than you'd think.

Tidal locking has to do with how two objects pull on each other. When the Moon first formed, it was much closer to the Earth than it is now. The Earth's gravity pulled on the Moon so much that it created a bit of a bulge on one side, and that bulge in turn pulled just a bit harder on the Earth than the rest of the Moon's surface did. Those two gravitational forces started amping each other up more and more, to the point that they now just constantly hold onto each other. As the Moon orbits the Earth, it rotates so that its bulgiest part is closest to the Earth.

While this tidal locking process was happening, the Moon and the Earth were both still super-heated from that initial impact. As the Moon's magma ocean cooled into a crust, it did so faster on the side facing away from the Earth's heat. As a result, the side facing the Earth formed a thinner crust. And it was almost entirely on this thinner side that magma was able to burst through the crust and form volcanoes.

This is why the far side of the Moon looks so different from the side that we see here on Earth. It lacks the dark splotches and has more impact craters. Both sides of the Moon would have experienced the same frequency of impacts, but on the volcanic side, flowing lava would have erased them or filled them in.

Now that we're on the topic of the far side of the Moon, I'd like to point out an important distinction. The 'far side' of the Moon is not the same as the 'dark side' of the Moon. A lot of people use these expressions interchangeably, but they're completely different concepts. The far side of the Moon is the part of the Moon that we never see from Earth. The dark side of the Moon is just whatever side of the Moon is facing away from the Sun, and that changes all the time as the Moon orbits the Earth.

There are parts of the Moon that are always dark, but aren't whole 'dark sides'. These parts are permanently shadowed craters near the poles, where sunlight never gets in. They are worth mentioning because they are unbelievably cold – colder than the surface of Pluto, which is about five billion kilometres farther from the Sun than the Moon is.

A POTENTIAL LAVA TUBE SKYLIGHT ON THE MOON, IMAGED BY THE LUNAR RECONNAISSANCE ORBITER SPACECRAFT. CREDIT: NASA/GSFC/ ARIZONA STATE UNIVERSITY

The reason these craters are so cold is that they're extremely deep, and their deepest parts get literally zero sunlight. And because there is no real atmosphere on the Moon to keep warmth circulating, zero sunlight means zero heat. On the surface of the Moon, the transition from day (facing the Sun) to night (facing away from the Sun) involves a temperature change of about 290 degrees Celsius, going from full-blast sunlight that hasn't been filtered through any atmosphere, to the unadulterated coldness of outer space. The cold is even more extreme in these deep craters because they haven't had any sunlight at all for billions of years.

THE FAR SIDE OF THE MOON WITH THE EARTH BEHIND IT. IMAGE TAKEN BY THE DEEP SPACE CLIMATE OBSERVATORY SPACECRAFT. CREDIT: NASA/NOAA.

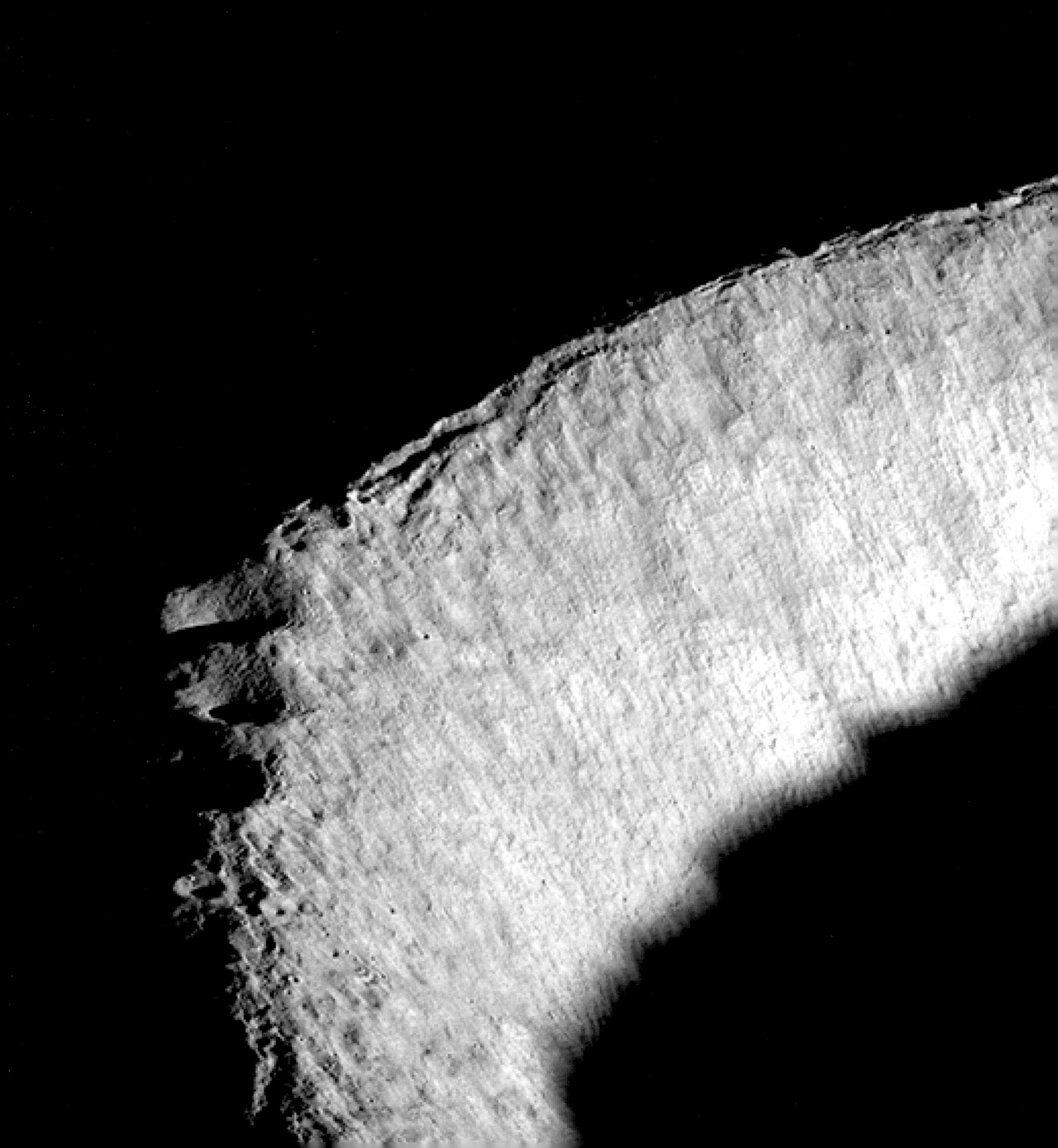

SHACKLETON CRATER, THE INTERIOR OF WHICH IS PERMANENTLY SHADOWED. IMAGE TAKEN BY THE LUNAR RECONNAISSANCE ORBITER SPACECRAFT. CREDIT: NASA/GSFC/ARIZONA STATE UNIVERSITY.

All this talk of light and darkness brings us to the phases of the Moon. You might already be familiar with terms such as new moon, full moon, crescent moon, waxing and waning. But the mechanics behind these phases are actually deeply, frustratingly complicated.

Here's the gist of it: the Earth goes around the Sun, the Moon goes around the Earth, and the Earth spins on its own axis. When we see the Moon rising and setting in the sky, we're actually the ones moving as the Earth spins beneath the Moon. The Moon takes about a month to go all the way around the Earth, and as it does this, it moves towards and away from the Sun. When the Moon is between the Earth and the Sun, its far side is lit by sunlight. We don't get to see that side of the Moon since it points away from the Earth, so we just see its dark side, which means we can't see it at all. This is what we call a new moon. When the Moon is on the opposite side of the Earth from the Sun, the side of it that we can see is totally illuminated by sunlight, so we get to see it in all its glory: a full moon. When it's halfway between these positions, we see half of it illuminated (called, counterintuitively, a quarter moon). And so on and so forth.

This all gets a bit complicated because things are tilted. You may remember from elementary school that Earth's tilt relative to the Sun is the reason why we have seasons. It's summer when the part of the world you live in is tilted towards the Sun, and it's winter when it is tilted away from the Sun. The Moon's orbit around the Earth is also tilted relative to the Earth (that is, it doesn't go around us directly above our equator), and relative to the Sun (that is, its path around Earth isn't parallel to Earth's path around the Sun).

SIDEBAR

THE UPSIDE-DOWN MOON

One of my all-time favourite Moon facts is that it looks upside down from the opposite side of the Earth. If you live in Australia and then travel to England, the Moon will look upside down compared to what you're used to seeing, and vice versa.

This is a slightly mind-bending thing that becomes a bit easier to comprehend if you start by imagining yourself looking at the Moon from one of the Earth's poles.

Because the Moon orbits the Earth around (though not directly in line with) the Earth's equator, when you are on either of the Earth's poles the Moon will never appear overhead; rather, it will always be relatively close to the horizon and you'll perceive the side closest to the ground as the bottom. Because a person standing on Earth's north pole is upside down compared to someone on the south pole, their perspective of the Moon would be upside down as well.

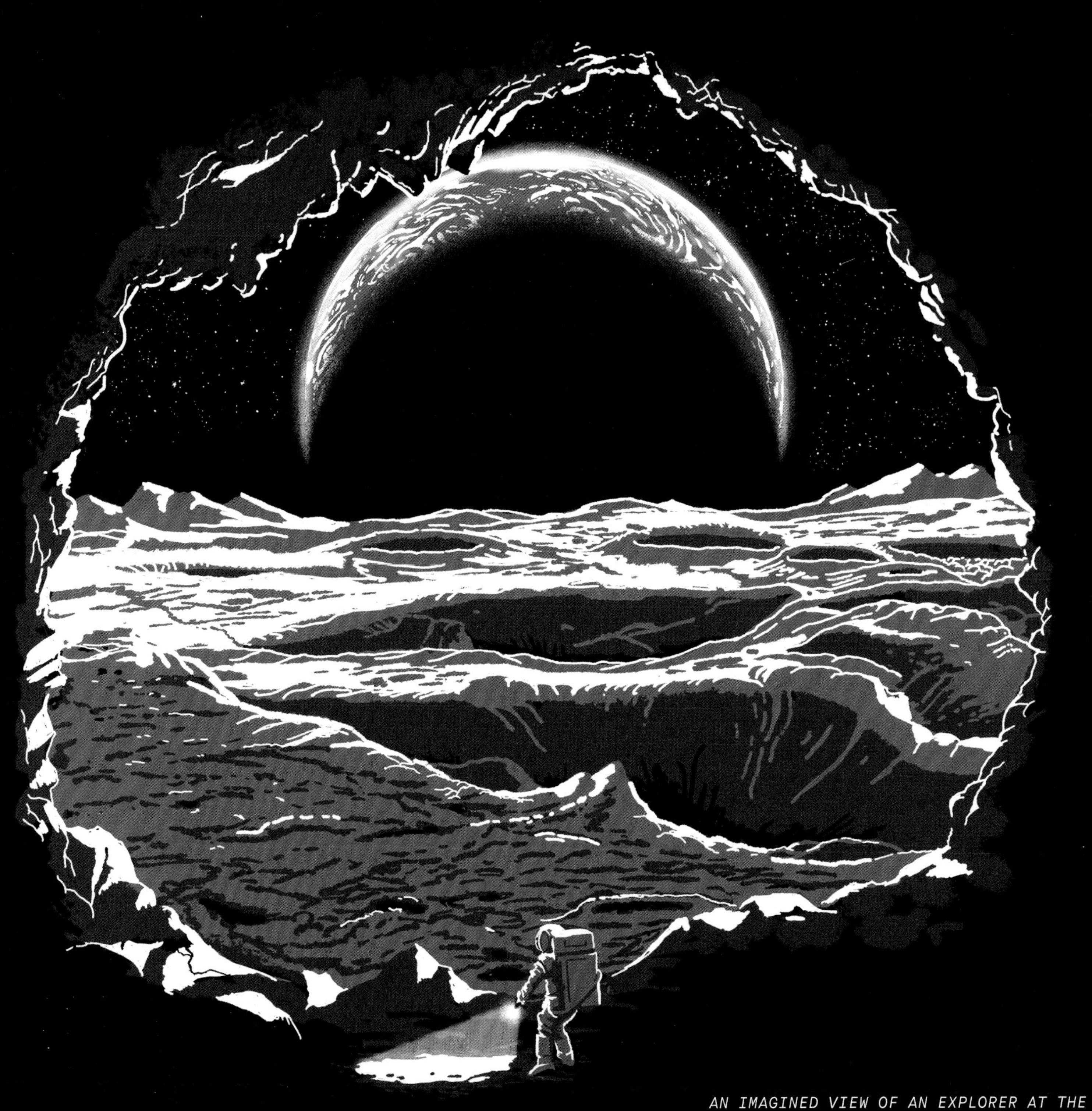

AN IMAGINED VIEW OF AN EXPLORER AT THE MOUTH OF A LUNAR LAVA TUBE. CREDIT: GORDON AULD.

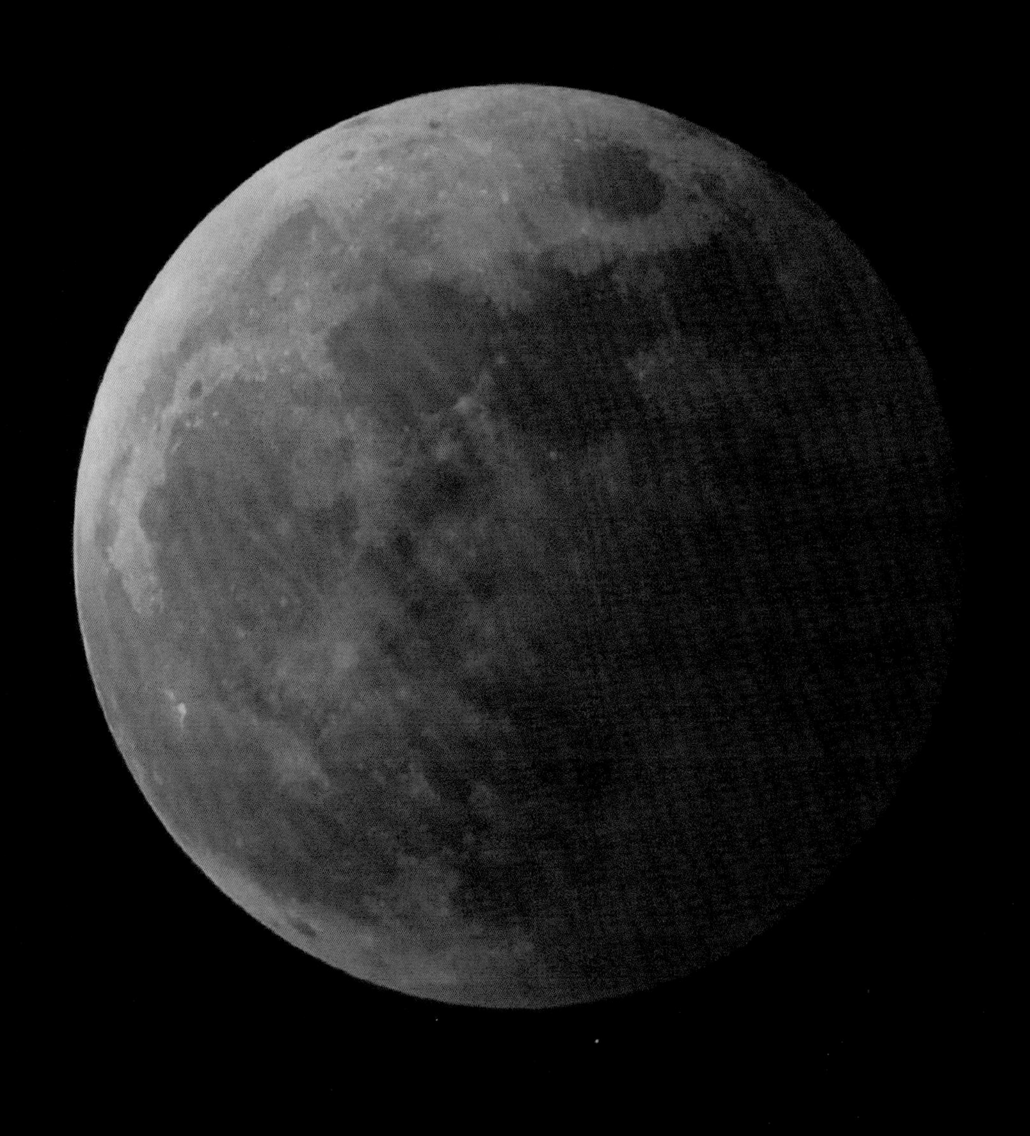

THE RED HUES OF A TOTAL LUNAR ECLIPSE.
CREDIT: NASA.

All this tilting makes things difficult to wrap your head around, and it's also why we don't get lunar or solar eclipses all the time. If everything were lined up perfectly, every time the Moon passed between the Earth and the Sun we'd get a solar eclipse, and every time the Moon was on the opposite side of the Earth we'd get a lunar eclipse. Instead, when the Moon is between the Earth and the Sun, it's normally either above or below the sightline between the two.

Every once in a while, though, the Moon's position in its tilty orbit around the tilty Earth does put it directly between us and the Sun, causing a total solar eclipse. And these eclipses are somewhat miraculous things.

Although I don't recommend looking directly at the Sun, you might have noticed that it looks to be about the same size in the sky as the Moon. This is a total coincidence and has to do with the size of the Sun and its distance from Earth, and the size of the Moon and its distance from Earth. Despite the Moon and Sun being wildly different in size, these measurements all line up to make both objects look to be the same. This means that when they are in the same spot in the sky, the Moon perfectly covers the Sun.

On many other planets, when a moon passes in front of the Sun, it only partly obscures it. But on Earth, thanks to this sheer coincidence of size and distance, when the Moon passes in front of the Sun, it obscures the sphere of the Sun but lets us see its outer atmosphere (called a corona, which means crown), resulting in a stunning solar eclipse.

Lunar eclipses are also spectacular to witness. These eclipses happen when the Moon passes completely into Earth's shadow, blocking any sunlight from directly illuminating the Moon. But this doesn't make the Moon completely dark – instead, it turns an eerie red colour. This occurs because some sunlight still reaches the Moon's surface after bouncing through the edges of Earth's atmosphere. Our atmosphere scatters blue light, which is why the sky looks blue during the day. At sunrise and sunset when the Sun is low on the horizon, the sunlight reaching you has passed through more atmosphere than it does when it's right overhead. This scatters away even more blue light, creating the red-tinged skies we see at dusk and dawn. The same thing happens during a lunar eclipse. Although the Moon is in shadow, the edges of Earth's atmosphere still scatter some red light onto the Moon's surface, giving it a reddish hue and inspiring the term 'blood moon'.

Another funky aspect of the Moon's orbit around the Earth is that it isn't perfectly circular. This means that at times the Moon gets closer to the Earth, and at other times, it's farther away.

This is even noticeable from Earth, making the Moon look a little bigger or smaller in the sky. When we get a full moon during the period when the Moon is closest, we call it a supermoon.

Although the size difference between a regular moon and a supermoon is technically detectable from Earth, it's unlikely that it's what you're seeing when you notice that the Moon looks particularly big. Instead, it's likely an optical illusion caused by a full moon rising or setting close to the horizon. Set against foreground objects like buildings or trees, the Moon can look startlingly big compared to when it's just hanging out in the sky with nothing but clouds for size comparison.

Overall, the Moon is super visible in the sky and changes regularly in ways that we kind of understand but that are honestly hard to wrap your head around. This combination is perfect fuel for spiritual belief systems, and people have forever been fascinated with the idea that the Moon can affect our behaviour. The words 'lunatic' and 'lunacy' come from the Latin *lunaticus*, meaning 'moonstruck', because even back in ancient Rome people thought that madness and epilepsy were caused by the Moon. Tales of werewolves are exaggerations of the common belief that people act strange during full moons. But where do these ideas come from?

My guess is that it's a pattern recognition bias – people look for patterns even where there aren't any, and people tend to see what they're looking for. When there is a big bright full moon once a month and weird things happen around that time, people are keen to blame it on that mysterious orb in the sky rather than accepting the randomness of the world. Basically, people remember the full moon nights when strange things happen, but forget all the full moon nights that are completely mundane.

The Moon does influence the Earth, even if it's not by making us crazy. Its gravity pulls on our planet, creating tides in the oceans and – more surprisingly – on the land. But it's not that surprising once you stop and think about it, because if the Moon is pulling on the oceans it's also pulling on the rest of the Earth. The oceans are more pliable, so they rise and fall by a greater amount. But the ground you're resting on now is also moving up and down as the Earth rotates.

THE MOON AS IT MIGHT BE SEEN FROM THE NORTHERN HEMISPHERE (LEFT) AND SOUTHERN HEMISPHERE (RIGHT) PERSPECTIVES. IMAGE CAPTURED BY THE LUNAR RECONNAISSANCE ORBITER SPACECRAFT AND ROTATED. CREDIT: NASA/GSFC/ARIZONA STATE UNIVERSITY/THE PLANETARY SOCIETY.

THE MOON, IMAGED BY THE GALILEO SPACECRAFT. CREDIT: NASA/JPL.

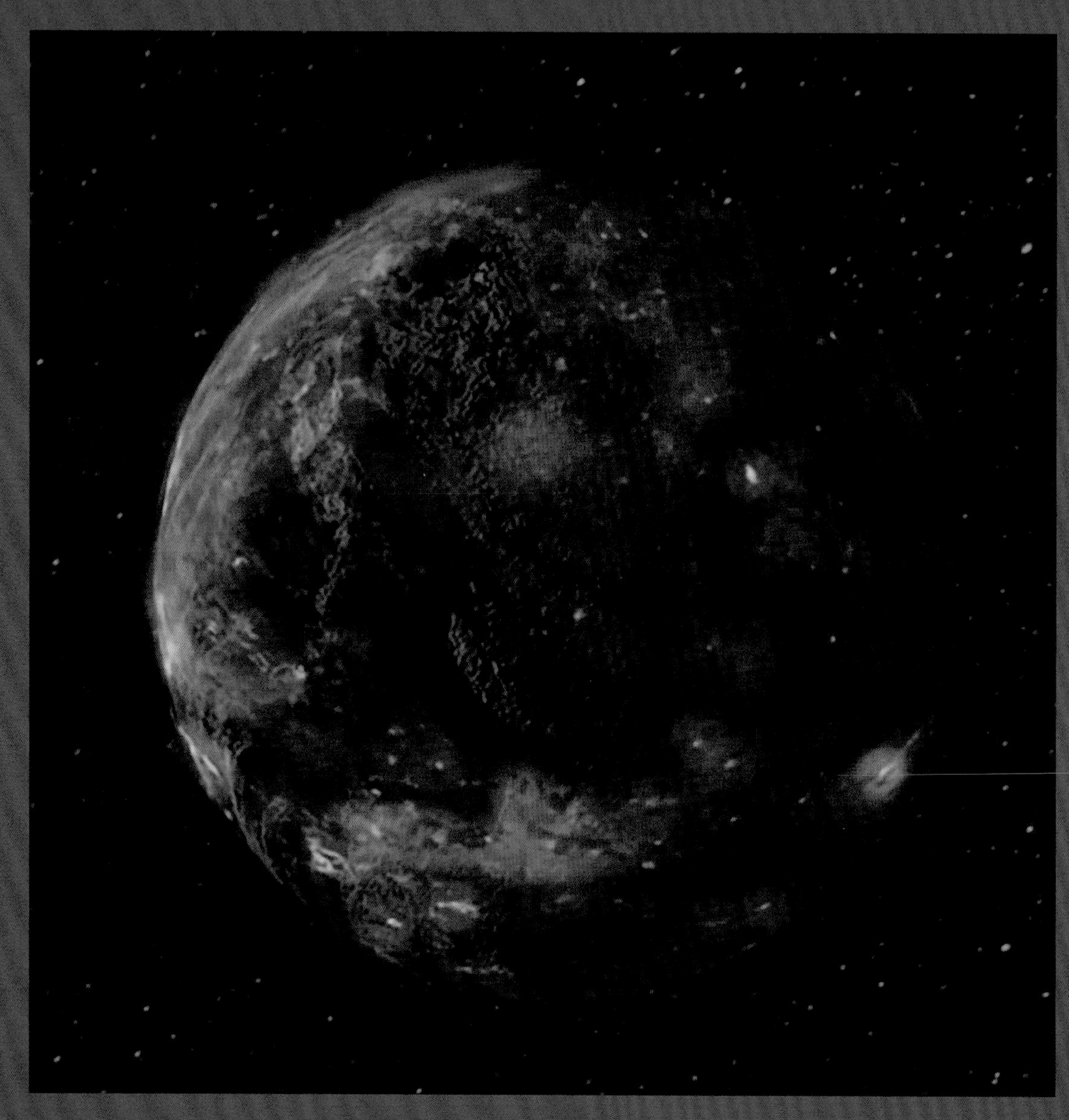

AN ARTIST'S IMPRESSION OF THE MOLTEN MOON. CREDIT: NASA.

The Moon's gravity also affects the Earth in another way. As the Moon pulls on the Earth, it actually slows down the Earth's rotation. It's a tiny effect, adding only two milliseconds to the Earth's rotation every 100 years, but in 2012, one second was added to the World Clock to account for this.

The Moon's pull makes every day a little bit longer. That means that today was literally the longest day of your life.

Another unique thing about the Moon is that we've actually been there. Twelve human beings have set foot on the Moon, all between 1968 and 1972.

These were all NASA astronauts, and since this was back in a less egalitarian time, this unfortunately meant that they were all white men. These days, the world's astronaut corps are a little more diverse.

The Moon-landing program has an important place in space history, and an interesting backstory. The six missions that brought those 12 men to the lunar surface were all part of the Apollo program, which was initiated by President John F. Kennedy in 1961 in the context of the Cold War. The US government was in a war of ideals with the Soviet Union, trying to show that capitalism was a superior economic system to communism. It was a massive struggle between two superpowers that both had nuclear weapons and both, thankfully, knew better than to use them. So instead of outright fighting each other, they competed in other ways. One of the arenas of competition was space. In 1957, the Soviets launched Sputnik, the first-ever orbital spacecraft, and this sparked a massive battle of space firsts:

★ November 1957: The Soviets put the first living animal in space (the dog Laika, RIP).

★ January 1959: The Soviets got a spacecraft out of Earth's orbit for the first time, sending Luna 1 on a fly-by of the Moon.

★ August 1959: The Americans took the first photo of Earth from space.

★ September 1959: The Soviets landed something on another celestial body (the Moon) for the first time.

★ October 1959: The Soviets took the first photo of the far side of the Moon (and, impressively, returned that spacecraft to Earth to retrieve the film).

★ August 1960: The Soviets managed to get an animal into space and back to Earth again without killing it – two animals, in fact: the dogs Belka and Strelka.

★ January 1961: The Americans sent the first ape into space, also bringing it back safely (Ham the chimpanzee).

★ April 1961: The Soviets sent the first human into space (Yuri Gagarin) (a big deal, obviously).

★ May 1961: The Soviets got a spacecraft to Venus for the first time, and soon thereafter did the same for Mars (June 1963).

★ June 1963: The Soviets sent a woman into space for the first time (Valentina Tereshkova) – notably, 20 years before the Americans sent a woman into space.

★ February 1966: The Soviets landed the first spacecraft on the Moon, and later on Venus (in March).

All of this happened fairly rapidly between 1957 and 1966, and it looked like the Soviets were getting all the best firsts. But throughout the 1960s, the US government invested ridiculous amounts of money in the newly formed National Aeronautics and Space Administration (NASA) – and it paid off. In 1968 the Apollo 8 mission took three astronauts into orbit around the Moon, and in 1969 Apollo 9 made Neil Armstrong and Buzz Aldrin the first humans to set foot on another celestial body. This (combined with some pretty significant problems going on in the Soviet Union) essentially won the space race. The Americans made it clear that they had the ability to do the seemingly impossible, and capitalism went on to rule the world – for better or for worse.

Five more Apollo missions landed astronauts on the lunar surface, and the astronauts that got to land did more than just take pictures, plant flags and play golf. They also collected 382 kilograms of lunar rocks to bring back to Earth for scientists to study, which has been instrumental to lunar research ever since. New discoveries are being made to this day as advances in research technology allow us to study these rocks in new ways.

By the time Apollo 15 made the fourth landing on the Moon, the US government was shelling out cash with abandon. The final three Apollo missions each included a Lunar Roving Vehicle, aka a moon buggy, a sort of stripped-down jeep that was useful for collecting rocks and allowing the astronauts to travel greater distances – and that was undoubtedly fun to drive.

The entire field of American space science owes a lot to the Apollo program. When the US government was pumping money into the space race, it established NASA centres across the country and hired thousands of highly skilled people. When the space race was won, it was politically impossible to shut down all those centres and lay off all those people. The politicians whose constituents worked for NASA had a huge political incentive to keep NASA well funded. The budget did get cut dramatically – which is why we don't have settlements on the Moon and Mars right now – but NASA was so embedded in the American political system that it remained disproportionately well funded.

To this day, NASA gets more funding than every other space agency in the world combined.

A lot of NASA's funding still goes to the astronaut program. Human spaceflight has some arguably useful spin-off applications in medical research and fun stuff like freeze-dried food, but overall it is not entirely about science. Luckily, plenty of money from that overall inflated pool of funding goes to space science missions and other important research, including climate research. And although it can be a hard pill for a leftist science nerd like me to swallow, without the Cold War we'd probably know a lot less about the universe than we do now.

THE EARTH RISING OVER THE LUNAR HORIZON, CAPTURED BY THE APOLLO 12 ASTRONAUTS. CREDIT: APOLLO 12 CREW/NASA.

SIDE BAR

DEFEATING MOON-LANDING DENIERS

In this crazy world of ours, there are people who hold some very strange and very wrong beliefs. One of those is that the Moon landings of the 1960s and 1970s never happened. The general idea is that the US government staged the Moon landings in an effort to beat the Soviets through pure propaganda.

Although I love a conspiracy theory as much as the next person, here are two extremely solid pieces of evidence that we definitely did land people on the Moon.

We have all that Moon rock they brought back. Unless the conspiracy stretches all the way to the lunar science community, who are faking their research to this day, the fact that we have hundreds of kilograms of lunar material brought back by the Apollo astronauts quells the argument that they never went.

We can see their landers, and even the moon buggy, from space. Spacecraft that are in orbit around the Moon take pictures of its surface, and in these photos we can see all the equipment that the astronauts used to get to and around the lunar surface.

FOR ALL MANKIND
CREDIT: NASA

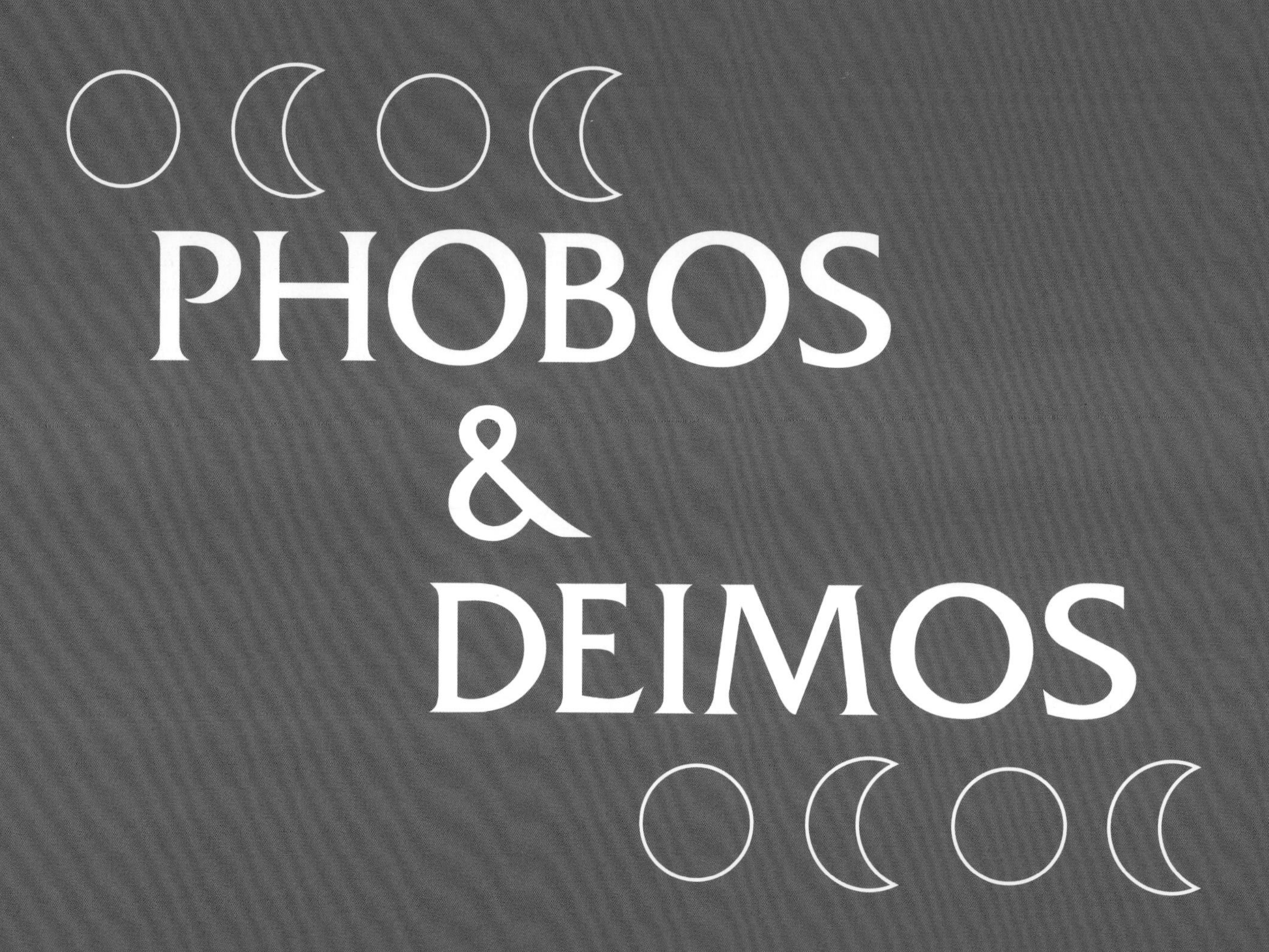

THE LUMPY MOONS

Mars is the only other rocky planet in the solar system with moons, and it only has two of them: Phobos and Deimos.

On first glance, these two moons don't look like much. However, they will likely be the first non-Earth-orbiting moons that humans are going to visit, simply because of their proximity to Mars – the planet humans are most obsessed with setting foot on.

Phobos and Deimos have a lot less going on than their freaky hellscape cousin Io or their icy ocean cousin Enceladus. This mostly comes down to their size, since the bigger a moon is, the more dynamic its internal processes are likely to be. Little moons like Phobos and Deimos don't have differentiated levels (such as a distinct crust, mantle or core), and these two are actually too small to even be fully round.

SIDE BAR

SIZE AND SHAPE

Round things in space like planets and large moons have this shape because of gravity. When something accumulates enough mass, the gravitational pull towards the centre of that object is very strong. That pull is exerted evenly in every direction, pulling all the stuff the object is made of into the centre equally and creating a round shape. The bigger something is, the stronger the gravitational effect and the rounder it becomes. More massive bodies are also less likely to have huge mountains for this same reason; gravity keeps them from rising too high. That's why smaller moons like Io have taller mountains than Earth.

What the object is made of also matters. Denser stuff like rock is harder to move, and requires more gravitational force to pull it into a sphere. Ice, which is less dense, can be more easily moulded into shape. This means that icy moons can more easily become spherical.

Io is the only round moon other than our Moon that is made of rock. There are other tiny rocky moons out there, but none other than those two are big enough to form into spheres.

The added factor of rotation makes planets and moons imperfectly spherical. As the object spins, mass is pulled outward in the direction it's spinning – sort of like how you get pulled outward when you're whipping around a merry-go-round. Most rounded planets and moons are a little fatter around the equator than from pole to pole.

Phobos and Deimos are way too small to get round. The smallest spherical thing in the solar system is Saturn's moon Mimas, which has a diameter about 17 times that of Phobos.

kilometres across. This size of moon is actually extremely common in the solar system. Giant planets like Jupiter and Saturn have a handful of big, noteworthy moons, but they also have dozens of small, lumpy moons like Phobos and Deimos.

Phobos and Deimos have some of the coolest names in the solar system. Mars is the Roman god of war, corresponding to the Greek god Ares. In the myths, this bloodthirsty god was usually accompanied by his sons Phobos, the personification of fear and panic, and Deimos, who embodied terror and dread. The trio was not to be trifled with.

Ironically, however, the Martian system is the one we humans trifle with the most. Mars is an extremely hot destination for space exploration, both human and robotic, because of its potential for life.

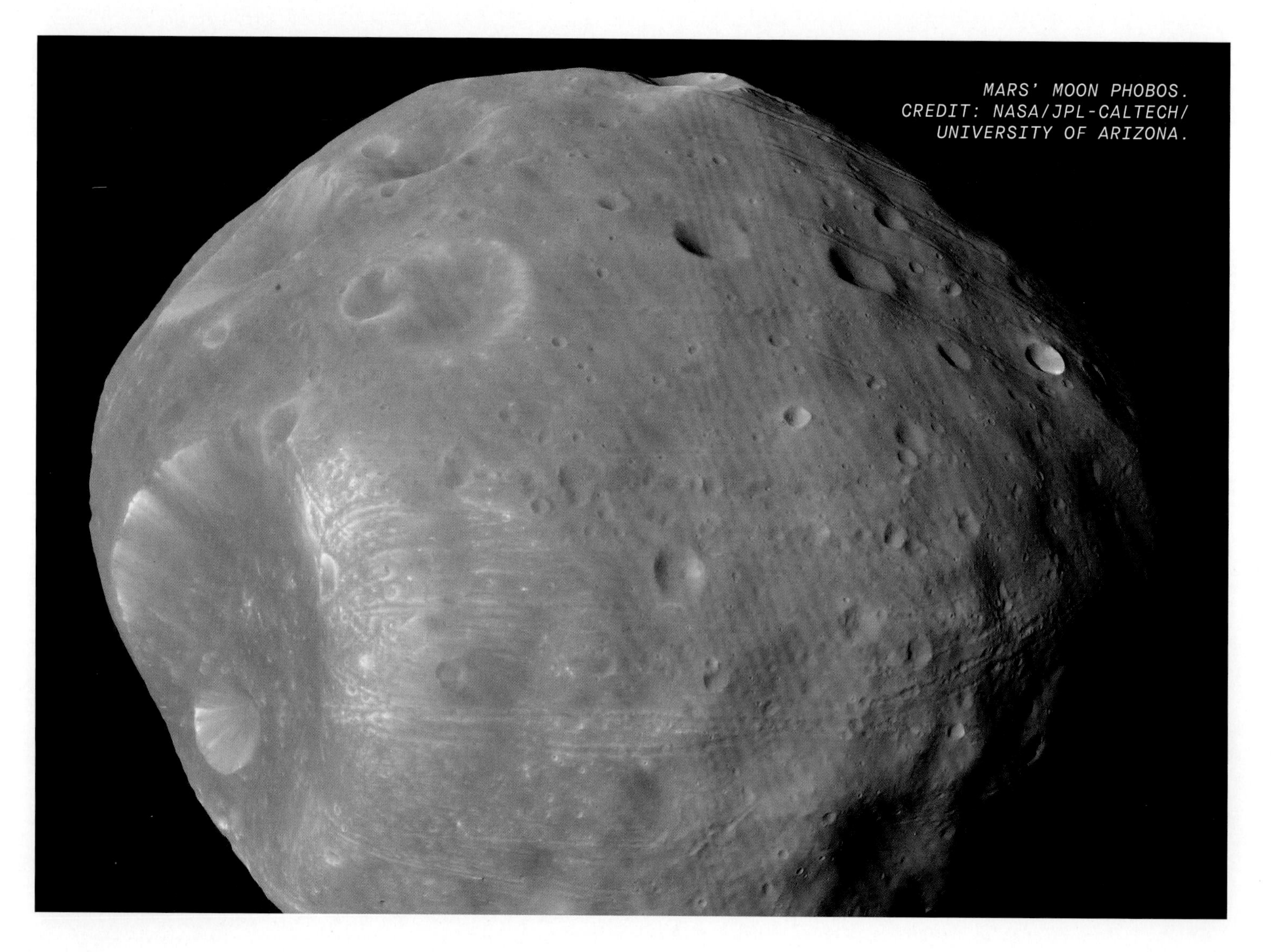
MARS' MOON PHOBOS. CREDIT: NASA/JPL-CALTECH/ UNIVERSITY OF ARIZONA.

SIDE BAR

WHY WE'RE OBSESSED WITH MARS

Humans have sent more spacecraft to study Mars than any other planet beyond Earth. It's also been named the 'ultimate destination' for human exploration by NASA. The main reason we're so fixated on exploring this particular world is that it used to be a lot like Earth, and therefore could have been home to life at some point in the past – and may be even today.

We've been interested in the red planet for a long time. Back in 1895, an astronomer used a telescope to look at Mars and thought he saw a series of canals crisscrossing the planet. He assumed these had to be structures built by Martian engineers. The idea that there were intelligent aliens on Mars was popular for decades, until our observing technologies got advanced enough to prove that the planet was, in fact, completely desolate.

Still, that didn't stop us. Mars is relatively easy to explore because it's right next to Earth in the planetary line-up. Since the early days of the space age, we've been sending spacecraft there and those missions have accumulated a lot of evidence that Mars used to be a warm, wet world, much like Earth.

The story that many orbiters, landers and rovers have pieced together over the last few decades is that Mars used to have a magnetic field that protected its atmosphere from solar radiation, allowing the planet to retain warmth and water. Mars had oceans, lakes, rivers – the watery works. But eventually, Mars' molten core cooled and hardened, and without that churning metal at its centre the planet lost its magnetic field. The Sun's powerful radiation stripped away the atmosphere, all the water evaporated, and the planet turned into the desiccated world we know today.

The big question is: when Mars was watery and pleasant, did life evolve on it like it did on Earth? Most of the robotic missions to Mars today focus on trying to answer that question, and any human missions to Mars would likely involve collecting tons of rocks to bring back to Earth to look for microbial fossils and the like.

One really cool idea is that life on Earth came from Mars, or that Earth life made its way to Mars. This hypothesis is called panspermia, and it involves rocks from one planet getting blasted into space by an asteroid impact and flying to the other world, with little microbial critters on board. This sounds far-fetched, but it is totally possible. Martian meteorites have been found on Earth, proving that rocks can get blasted from one planet to another. There are also microorganisms that have proven their ability to survive in the vacuum of space for extended periods of time.

Phobos is the prime option for a Mars base, since it's the closer of the two moons.

Phobos is actually closer to its planet than any other moon in the solar system.

It only gets two centimetres closer per year, but that adds up; in 30 to 50 million years it is expected to get close enough to Mars for the planet's gravity to rip it apart, making it either crash into Mars or form a ring. In fact, there are a few strings of craters on the Martian surface that are likely the result of other small moons suffering this same fate. Basically, we can't take forever if we want to establish a human base on Phobos.

Phobos is already so close to Mars that when an asteroid hits the planet, the debris from the collision hits Phobos, creating streaks on its leading side (the side facing the direction it's moving through space).

An alternate theory for these streaks is that they came from a mega-impact that would have almost completely destroyed Phobos. The moon has a nine-kilometre-wide impact crater, which is huge for a moon its size. This impact would have been especially destructive because Phobos is far from dense. It's more a pile of rubble held together by a thin crust, with about a third of the entire moon made up of empty space. This extremely low density means that Phobos also has a very low mass, and accordingly weak gravity. A 150-pound person would only weigh two ounces on the surface of Phobos.

Continuing with this imagination exercise, if you were to stand on the surface of Phobos, you would have a phenomenal view. From the surface of the moon, Mars would look 6400 times larger and 2500 times brighter than the full moon appears from Earth, and it would take up a quarter of the width of your entire view of the sky. And because Phobos' rotation keeps the same side facing Mars at all times, you would have an enormous view of Mars all day every day.

Phobos might also be a useful home base for Mars exploration because it's much easier to land on and take off from than the planet itself. Anything that lands on Mars would hurtle down at high speeds because of the planet's gravity and its lack of a dense atmosphere to help slow things down. Getting materials to the Martian surface to use for building habitats would therefore be extremely tricky.

An easier tactic would be to have your human base on Phobos. Astronauts could then make little forays down to the Martian surface to plant flags, collect rocks, play golf, and whatever else astronauts do. The astronauts on Mars could also get a cool view when looking back up at their home base, since Phobos is close enough to have a dramatic effect when it passes in front of the Sun.

At about half the size of Phobos and almost four times farther from Mars, Deimos makes for a less practical home base. It is so small and has such low gravity that if you were on its surface and rode a bike off a small jump (as one does in space), you could get enough air to escape the moon's gravity and fly off into space.

DEIMOS OVER MARS. CREDIT: EMIRATES MARS MISSION.

TITAN

EARTH'S FROZEN SISTER

Saturn's moon Titan is completely unlike any other moon and feels more like an alien planet from a work of science fiction – it is also my favourite moon.

Titan is the second-largest moon in the solar system, and it is larger (although not more massive) than the planet Mercury. It's the only moon with a dense atmosphere, which stretches 600 kilometres high and completely obscures Titan's surface. And what makes Titan truly remarkable is that it is the only place in the solar system other than Earth with liquid lakes, rivers and seas on its surface.

Now, before you get too excited (and understandably so after hearing me go on and on about how crucial water is in the search for life), I need to clarify: Titan doesn't have seas of water – but of methane and ethane.

On Earth, we know methane and ethane as common natural gases. Methane in particular is very common as a fuel source – you might even have it pumped into your house for your gas oven or water heater. Methane is also one of the big greenhouse gases, and is known for being released into the atmosphere in the belches of pigs, cows and other livestock.

Just like water can go from gas to liquid to ice, methane and ethane change form according to the temperature.

The average surface temperature on Titan is about -180 degrees Celsius, cold enough for water ice to be as hard as granite. At this temperature, methane and ethane take their liquid forms and coalesce into lakes and even huge seas. These are the only stable bodies of surface liquid in the solar system other than on Earth.

Scientists had speculated that Titan could potentially have liquid methane and ethane on its surface long before they found concrete evidence of it. They saw data and images from Voyager 1 and 2's fly-bys of Titan that showed a dense atmosphere, and calculated that at those temperatures and with that kind of atmosphere, liquid hydrocarbons like methane and ethane could be possible. When other missions including the Hubble Space Telescope and the Cassini Saturn orbiter got a better look at Titan using radar and infrared filters, we received confirmation of the existence of these lakes and seas.

Titan gets its beautiful sea-speckled map thanks to its thick atmosphere. In fact, it is the only moon in the solar system with a dense atmosphere, which is even thicker than most planets'. Four times as thick as Earth's, Titan's atmosphere is so dense that you can't see through it at all with optical telescopes. Even sunlight can barely get through. Being a few billion kilometres farther away from the Sun, Titan gets about one per cent as much sunlight as Earth, and only ten per cent of that one per cent gets through the atmosphere to the surface. (This still provides decent light, surprisingly.)

If you were standing on the surface of Titan, it would look about a thousand times dimmer than daytime on Earth, but that's still five hundred times brighter than a fully moonlit night.

Surprisingly, this little moon has managed to hold onto its atmosphere despite not having a magnetic field of its own to protect it. Remember, Mars lost all its water when its core stopped generating a magnetic field and the Sun's radiation was able to blast its surface, stripping away its atmosphere. Although Titan doesn't have its own radiation-repelling magnetosphere, it orbits within Saturn's extremely large and powerful magnetic field. Saturn essentially shields Titan from solar radiation, allowing it to maintain an atmosphere.

You might be wondering why Jupiter's Titan-sized moons don't have atmospheres too, considering Jupiter also has a super-powerful magnetic field. Scientists think it comes down to the conditions in which the Jupiter and Saturn systems formed.

When Jupiter was in the process of coalescing into a planet, it was extremely hot. This heat and energy would have driven most of the gas out of the interiors of the moons forming around the baby planet. Saturn, on the other hand, formed under colder conditions. This let Titan form with gases trapped within its icy interior,

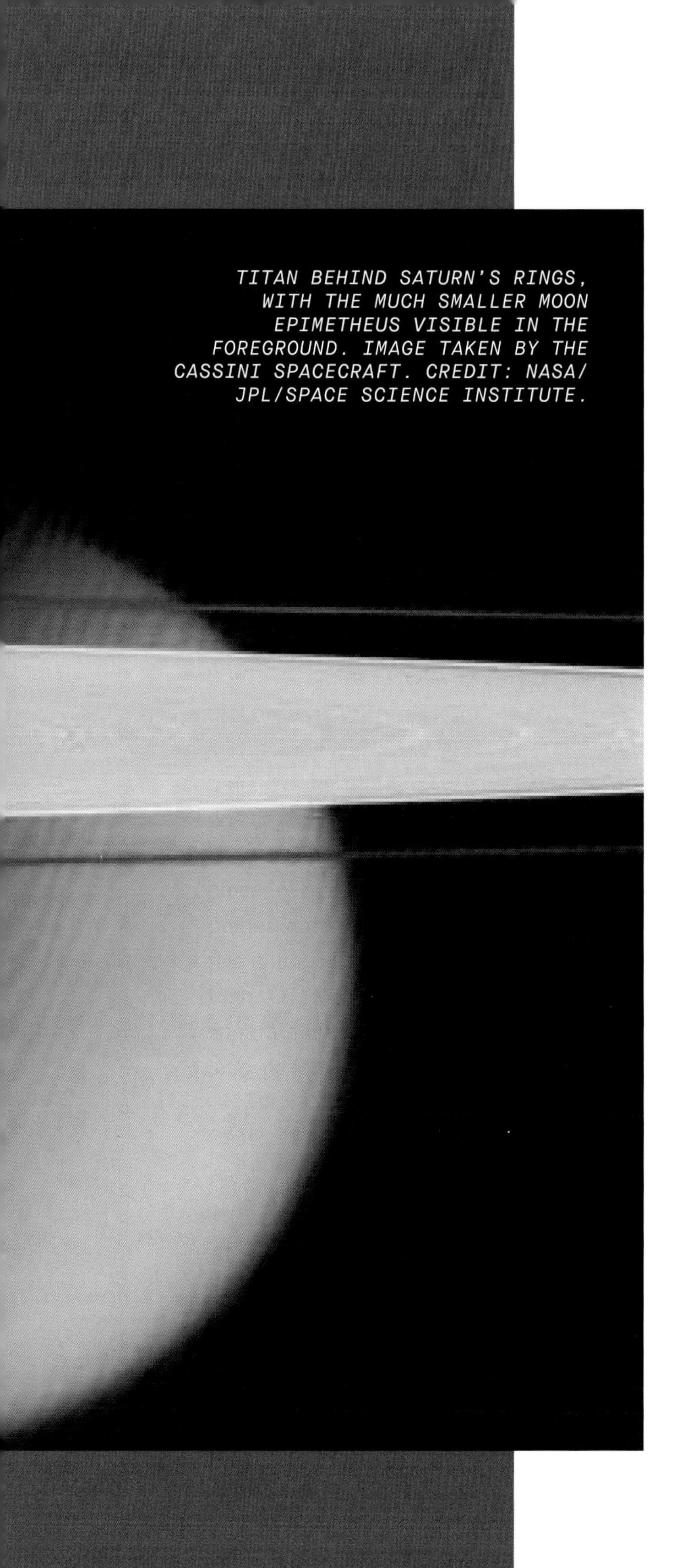

TITAN BEHIND SATURN'S RINGS, WITH THE MUCH SMALLER MOON EPIMETHEUS VISIBLE IN THE FOREGROUND. IMAGE TAKEN BY THE CASSINI SPACECRAFT. CREDIT: NASA/JPL/SPACE SCIENCE INSTITUTE.

which are gradually being released to the surface to form its atmosphere.

We know that Titan is probably still disgorging gases into its atmosphere because it contains a lot of methane. Most of Titan's atmosphere is nitrogen, like Earth's, but it also has a lot of methane that gives the moon its orange, smoggy appearance. Methane tends to be broken up by the Sun's ultraviolet light, which gets through Saturn's protective magnetic field, so Titan would have lost all its atmospheric methane long ago if it didn't have a constant supply of new methane from somewhere inside the moon. One theory is that there are cryovolcanoes on Titan that spew out methane gas along with water and ammonia.

The atmosphere's ability to block sunlight helps Titan to keep the liquid on its surface from evaporating (which liquids in space seem to really want to do). What's more, the pressure that all that gas exerts on the surface also helps to keep the liquid stable, by basically holding it down.

Because of this atmosphere, Titan has areas on its surface that look more like Earth than anywhere else in the solar system. Maps of Titan look like something that have come straight out of a fantasy author's imagination. And, fittingly, mountains on Titan are named after mountains in J. R. R. Tolkien's *Lord of the Rings* books, and plains and canyon systems on Titan are named after planets from Frank Herbert's *Dune* series. (Lest we forget that scientists are, fundamentally, huge nerds.)

Most of Titan's lakes and seas with their connecting rivers are concentrated near the poles, where there is even less sunlight capable of evaporating the tenuous liquid. There are, however, a few lakes closer to the equator, which scientists think are supplied by underground aquifers like oases in Earth's deserts.

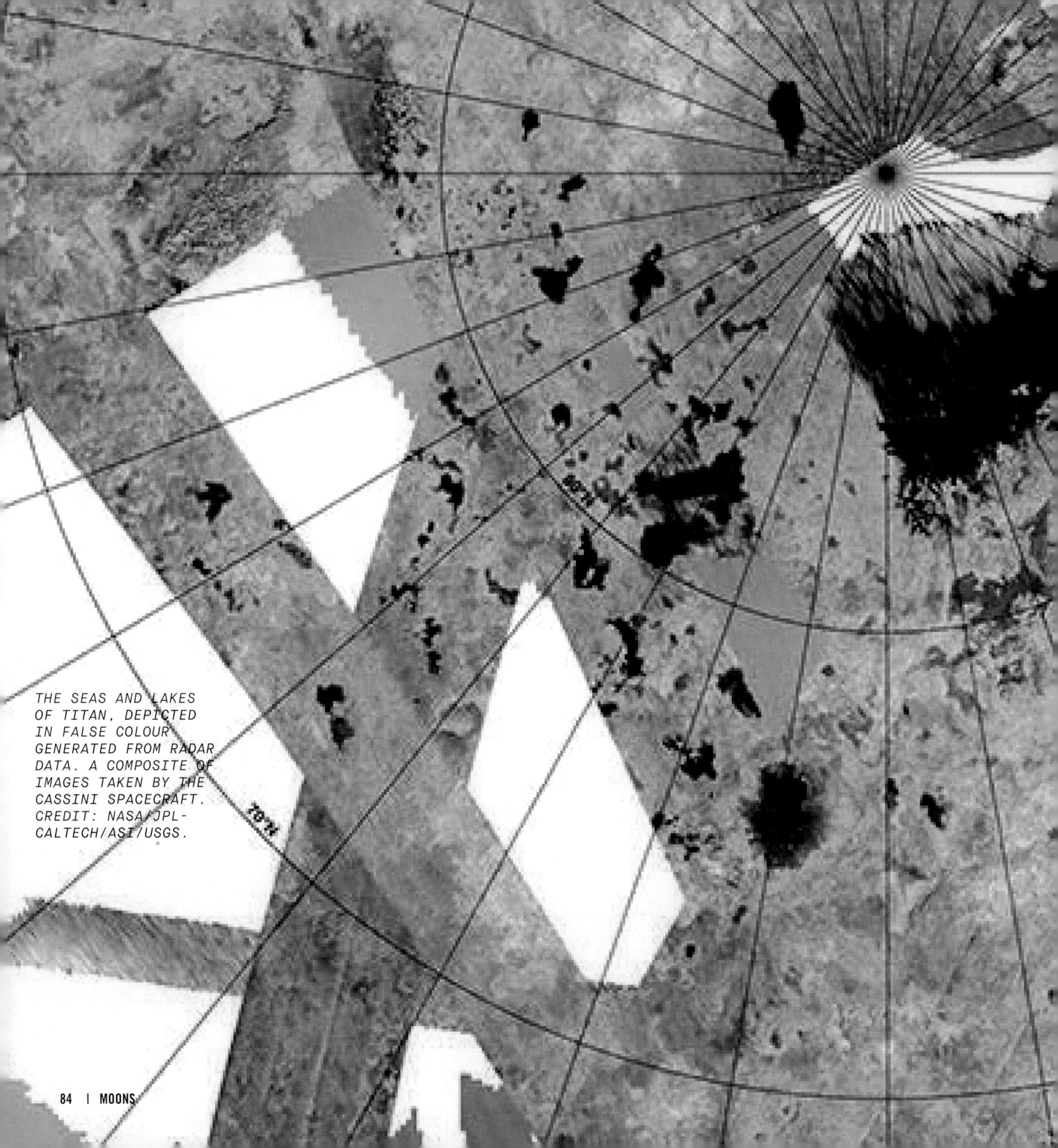

THE SEAS AND LAKES OF TITAN, DEPICTED IN FALSE COLOUR GENERATED FROM RADAR DATA. A COMPOSITE OF IMAGES TAKEN BY THE CASSINI SPACECRAFT. CREDIT: NASA/JPL-CALTECH/ASI/USGS.

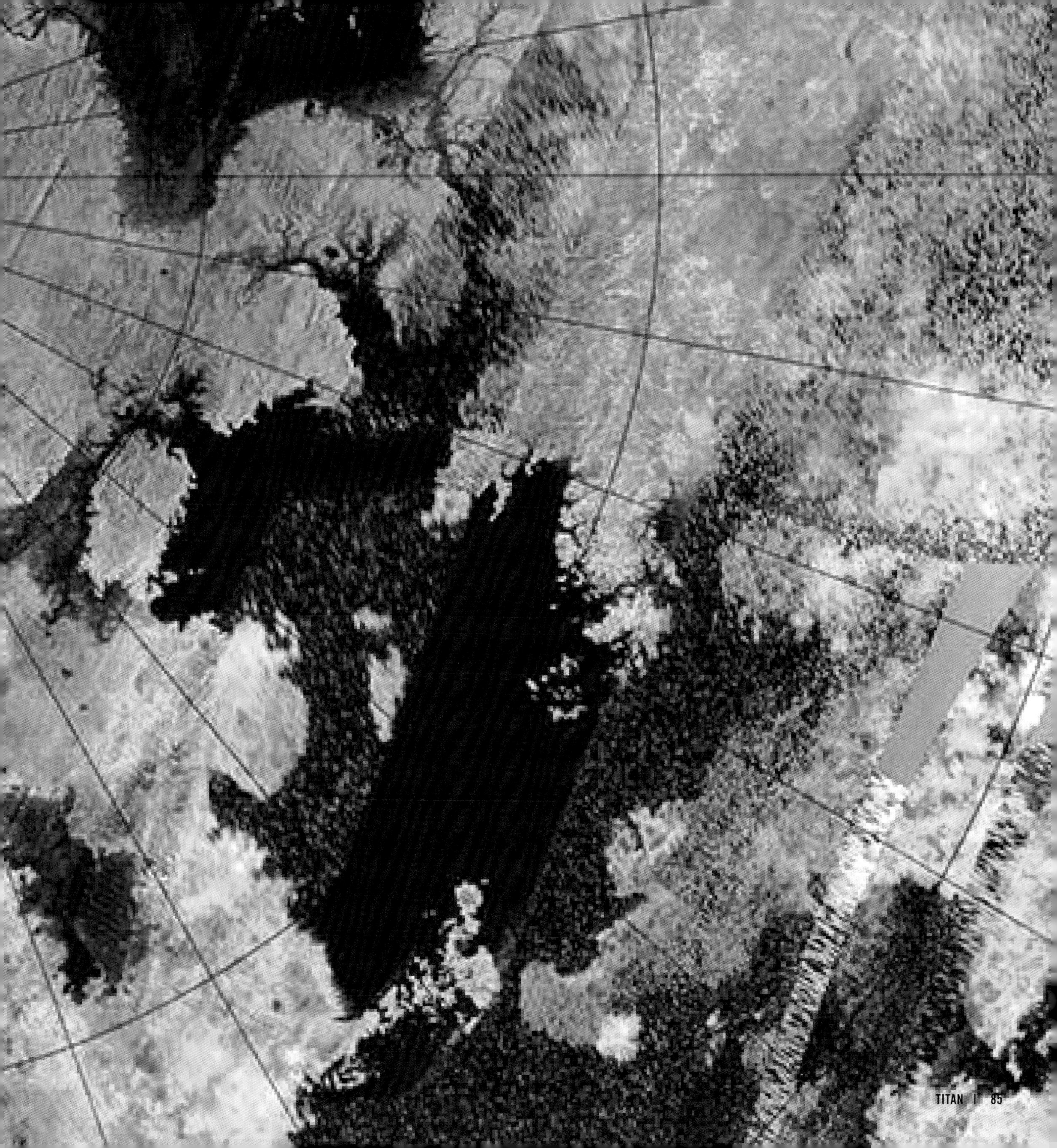

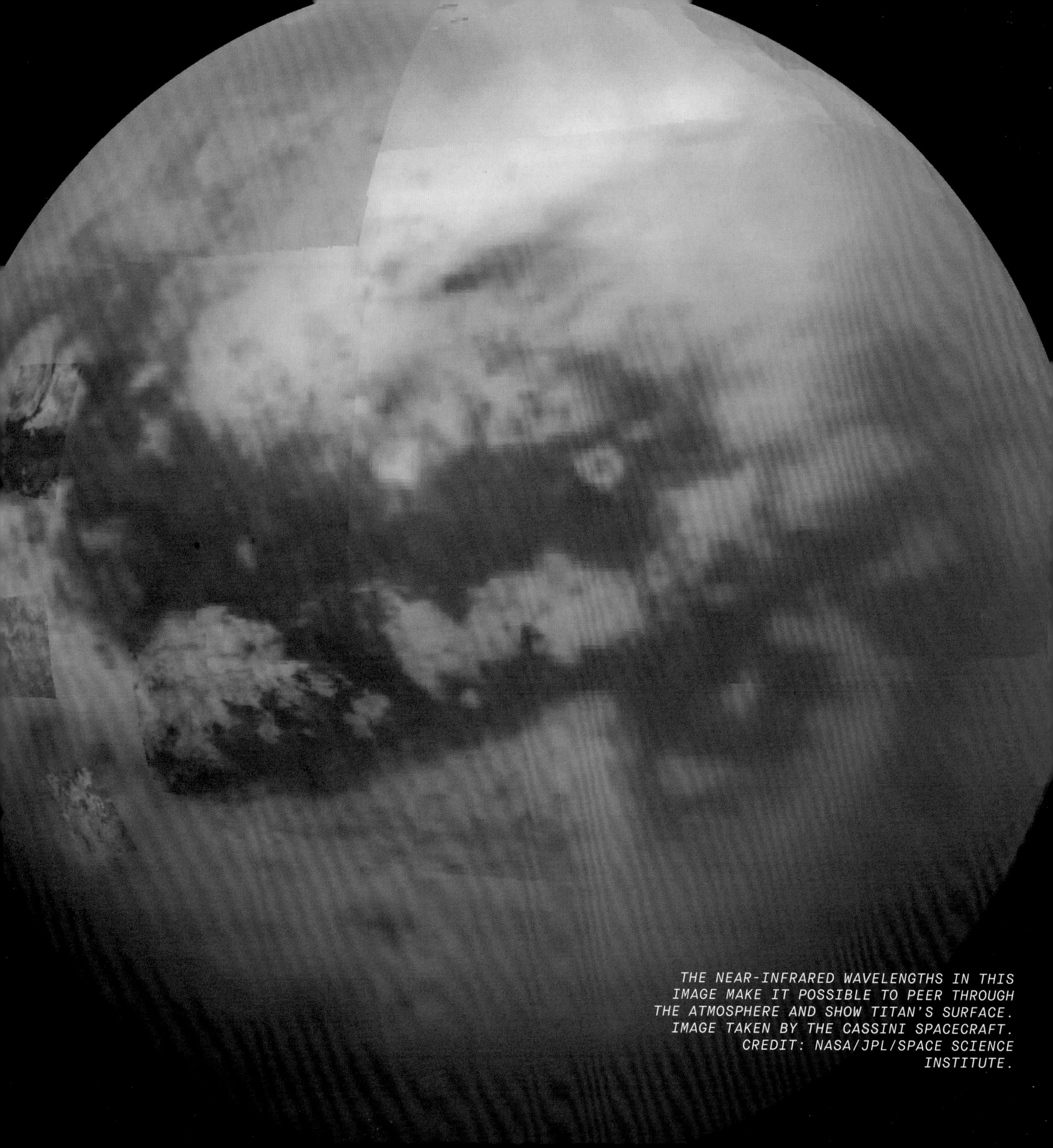

THE NEAR-INFRARED WAVELENGTHS IN THIS IMAGE MAKE IT POSSIBLE TO PEER THROUGH THE ATMOSPHERE AND SHOW TITAN'S SURFACE. IMAGE TAKEN BY THE CASSINI SPACECRAFT. CREDIT: NASA/JPL/SPACE SCIENCE INSTITUTE.

SIDEBAR

TITAN'S DARK PLAINS

For several years I thought that the dark areas in images like this one were Titan's lakes and seas - I was wrong. The large bodies of liquid methane and ethane on Titan are closer to the poles, and these dark splotches in infared images of the surface are actually huge plains.

These dark plains are interesting in their own right. Titan's surface consists of high-altitude areas of water ice that reflect a lot of light, and lower, darker regions. Scientists think that these dark areas used to be seas of liquid hydrocarbons that have since dried up. Now, they're plains made up of sand dunes, with sand unlike anything we see on Earth.

Here on our home planet, sand is made of grain-sized particles that were once part of something bigger that has been broken down into smaller pieces, usually by the motion of water. That original, bigger stuff differs from place to place - black sand from volcanic rock, pink sand from corals, and the more common whitish/ yellowish sand from silicate rock.

In Titan's equatorial dunes, you get sand made of hydrocarbons - the same kind of compounds that fill its seas in liquid form and its atmosphere in gaseous form. Rather than coming from big stuff that has been worn down, Titan's sand grains grow from smaller stuff sticking together. Scientists think that tiny collections of hydrocarbons the size of smoke particles (that is, very, very small) fall from Titan's sky and, once on the ground, stick together to form larger grains. These grains are about the size and consistency of coffee grounds, and they form huge sand dunes up to 100 metres high and tens to hundreds of kilometres long.

Along with hundreds of smaller lakes, Titan has three large seas about the size of North America's Great Lakes. (Terminology note: on Earth, seas can either be freshwater bodies that drain into the ocean, or saltwater bodies that are partly surrounded by land. On Titan, seas are defined as particularly large lakes.)

Kraken Mare, whose name comes from the mythological Norse sea monster the Kraken and the Latin word *mare*, which means 'sea', is the largest of the seas. It has an area of about 500,000 square kilometres, bigger than the Caspian Sea on Earth and larger than any other lake in the solar system. Its depth has been measured at 100 metres, but some data suggest it could be more than 300 metres deep in some places. Kraken Mare's composition seems to be mostly methane, with a little ethane and some other hydrocarbons mixed in.

Nearby is Ligeia Mare, named after one of the Sirens of Greek mythology. At about 126,000 square kilometres, it is a little bigger in surface area than Earth's Lake Superior. Like Kraken Mare, Ligeia is mostly made of liquid methane – enough to fill three Lake Michigans.

Punga Mare, the smallest of the seas, is named after the Māori mythological figure who was a son of Tangaroa, the god of the sea, and who was the ancestor of sharks, rays and lizards. Punga Mare, which is almost at Titan's north pole, stretches about 60,000 square kilometres. Unlike its larger cousins, Punga is a mix of ethane and methane.

Scientists still don't know why the composition of Titan's various lakes and seas differs, but one possibility has to do with Titan's other eerily Earth-like feature: precipitation.

Here on Earth we're familiar with the water cycle: water warms up and evaporates into the atmosphere, where it condensates in clouds, rains back down to Earth, collects into streams and rivers, flows down into lakes, gets evaporated again, and on and on.

THE SUN GLINTING OFF TITAN'S NORTH POLAR SEAS. IMAGE TAKEN BY THE CASSINI SPACECRAFT. CREDIT: NASA/ JPL-CALTECH/UNIVERSITY OF ARIZONA/UNIVERSITY OF IDAHO.

Titan is the only other place in the solar system with the same kind of cycle – only it's with liquid methane. On Titan, you get methane rainstorms, rivulets of methane running down mountainsides, huge methane rivers carving down the valleys that they pass through – just about everything rain does on Earth, methane does on Titan. This includes concentrating in certain areas, which might explain why some lakes have more methane in them than others.

Although Earth and Titan both have liquids that follow the same cycles, the differences between the two worlds make for very different behaviour in the liquids. On Titan, the combination of low gravity and a dense atmosphere causes rain to fall more slowly than it does on Earth. Here on our planet, rain falls at about 9.2 metres per second; on Titan, scientists think it probably falls at a fifth that speed, at about 1.6 metres per second. Likewise, the conditions on Titan can sustain raindrops about 1.5 times the size of the largest we get on Earth.

Trying to swim in Titan's lakes would also be very different from taking a dip in Earth's lakes.

Beyond the obvious factors of extreme cold and lack of oxygen, you wouldn't be able to enjoy a swim in Kraken Mare because methane isn't dense enough to hold you up.

Human bodies have a similar density to water, which is why we can float without having to paddle too much. A rock, on the other hand, is a lot denser than water which is why it sinks immediately. Liquid methane is about half as dense as water, so it wouldn't be able to hold us up. If you did a cannonball off the dock into a Titanian lake, you'd sink like an actual cannonball.

Despite being no fun for human swimmers, Titan's bodies of liquid methane could potentially be habitable to alien life forms. Astrobiologists (scientists who study the possibility of life beyond Earth) have proposed the possibility of organisms that breathe in hydrogen instead of oxygen, metabolise it with acetylene instead of glucose, and breathe out methane instead of carbon dioxide. Instead of using water as a solvent, Titanian life forms could use liquid methane. It's all theoretically possible, and some scientists have argued that there might even be evidence that it's happening. Hydrogen levels are lower at the surface than they are in the upper atmosphere. Likewise, acetylene levels at the surface are lower than expected. These effects could conceivably be due to some kind of alien life, but astrobiologists acknowledge that other explanations are more likely. Still, the possibilities are out there and the only way to find out more is by exploring Titan further.

AN IMAGINED VIEW OF ASTRONAUTS LOOKING OUT ACROSS TITAN'S DUNES. CREDIT: GORDON AULD.

A NEAR-INFRARED VIEW OF TITAN'S ATMOSPHERE. IMAGE TAKEN BY THE CASSINI SPACECRAFT. CREDIT: NASA/JPL/SPACE SCIENCE INSTITUTE.

TITAN'S UPPER ATMOSPHERE. IMAGE TAKEN BY THE CASSINI SPACECRAFT. CREDIT: NASA/JPL/SPACE SCIENCE INSTITUTE.

That being said, Titan is one of the better-explored moons of the outer solar system, and is actually the only moon other than Earth's Moon on which a spacecraft has ever landed.

NASA's Cassini spacecraft, which orbited Saturn from 2004 to 2017, was accompanied by a small lander provided by the European Space Agency. The Huygens lander, named after the Dutch astronomer Christiaan Huygens who studied Saturn's rings in the 1600s, hitched a ride to the Saturn system attached to the Cassini spacecraft. A few months after Cassini entered into orbit around Saturn, it passed close to Titan and dropped off the Huygens probe.

Huygens coasted towards Titan for 22 days. To conserve its small batteries, it kept its onboard computers dormant until a wake-up timer went off. Then, 15 minutes later, it entered Titan's atmosphere and began a two-and-a-half-hour descent. As soon as it had slowed down enough to blast off its heat shield, Huygens started taking photos and collecting data about the moon's atmosphere.

Most of the time, when we send a spacecraft to land on another world, we meticulously plan where it's going to touch down. We scope out the planet or moon using orbiters and pick a spot that's of scientific interest, with conditions that will make landing easiest. But with Titan, none of this was possible. We just dropped the lander on a trajectory that would take it to the moon and hoped for the best. The mission team had to prepare for the possibility that Huygens might land in a methane lake, on rock-hard ice, or somewhere else altogether.

Luckily, the landing site wound up being unproblematic. Using three different parachutes, Huygens slowed down enough that when it hit the ground it was only going about as fast as a ball being dropped from a metre high on Earth, bouncing and wobbling a little and avoiding any damage. It touched down on a relatively flat plain with a cushioned surface, about the consistency of packed snow and covered in small rocks and pebbles of ice.

THE HUYGENS PROBE BEING INSTALLED IN ITS DESCENT MODULE. CREDIT: ESA.

Huygens' batteries only lasted for three hours, and it could only send data back to Earth via Cassini, which would only be overhead for about that amount of time too. Within that very limited window, Huygens managed to learn a huge amount about Titan.

As it came down through Titan's atmosphere, Huygens identified the gases it was composed of, the temperature at different elevations, the movement of winds and how the pressure changed as it fell. The images it took from above the surface showed networks of rivers feeding into larger ones, canyons and valleys eroded by rains, and signs of past floods and upwelling from groundwater (or rather, groundmethane). At its landing site, Huygens found signs of a dried lakebed, where the rocks had been rounded by flowing liquid at some point in the past. And like so many other missions, the discoveries Huygens made sparked even more questions that scientists today are still working on answering.

Luckily, there is another mission to Titan in the works. NASA's Dragonfly mission will send a spacecraft to Titan, aiming to launch from Earth in 2027 and arrive on the Saturnian moon in 2034.

But this is no ordinary lander or orbiter; Dragonfly is a drone-like spacecraft with eight rotors (aka an octocopter) that will fly around Titan. The spacecraft is designed for a 2.7-year baseline mission, but it's very likely that it will outlast that conservative timeline. The octocopter will fly from site to site to study Titan's atmosphere, geology and chemistry, looking for any evidence that Titan might once have been habitable to microbial life, or even have life on its surface today.

Scientists have already picked a landing site and are timing the launch so that Dragonfly lands during a time of year when Titan's weather is particularly calm in that area. The octocopter is expected to fly a total of more than 175 kilometres, which is almost double the distance that all the Mars rovers have travelled combined.

Part of the reason Dragonfly is going to be able to fly so far is that it's really easy to fly in Titan's atmosphere. The combination of low gravity and dense air means that it doesn't take much to get off the ground. In fact, Titan is the easiest place in the solar system for flying. The only other worlds where there is enough atmosphere for flight (not including blasting straight up with a rocket) are Venus, Mars, Earth, Jupiter, Saturn, Uranus and Neptune. And all of those have much higher gravity than Titan, making it harder to take off. On Titan, which has a sixth of Earth's gravity and 1.5 times its atmospheric density, it could even be possible for a human (well protected by a spacesuit, of course) to strap wings onto their arms, flap those wings and take off.

And boy, would it be awesome to do that. Aside from being one of the more scientifically intriguing worlds of the solar system, Titan is also the place I'd most want to visit. Imagine standing on a huge dune of coffee grounds, watching chubby raindrops falling slowly from the hazy skies above you, looking out over a huge lake of weirdly alien liquid, or, if you had some kind of atmosphere-penetrating goggles, looking up to see Saturn and its glorious rings shining above you, looking ten times bigger than the Moon does from Earth. Then imagine flapping your strap-on wings and soaring off. This is the stuff of sci-fi fantasy, but what makes it all the more awesome is that it's in the realm of the real and possible.

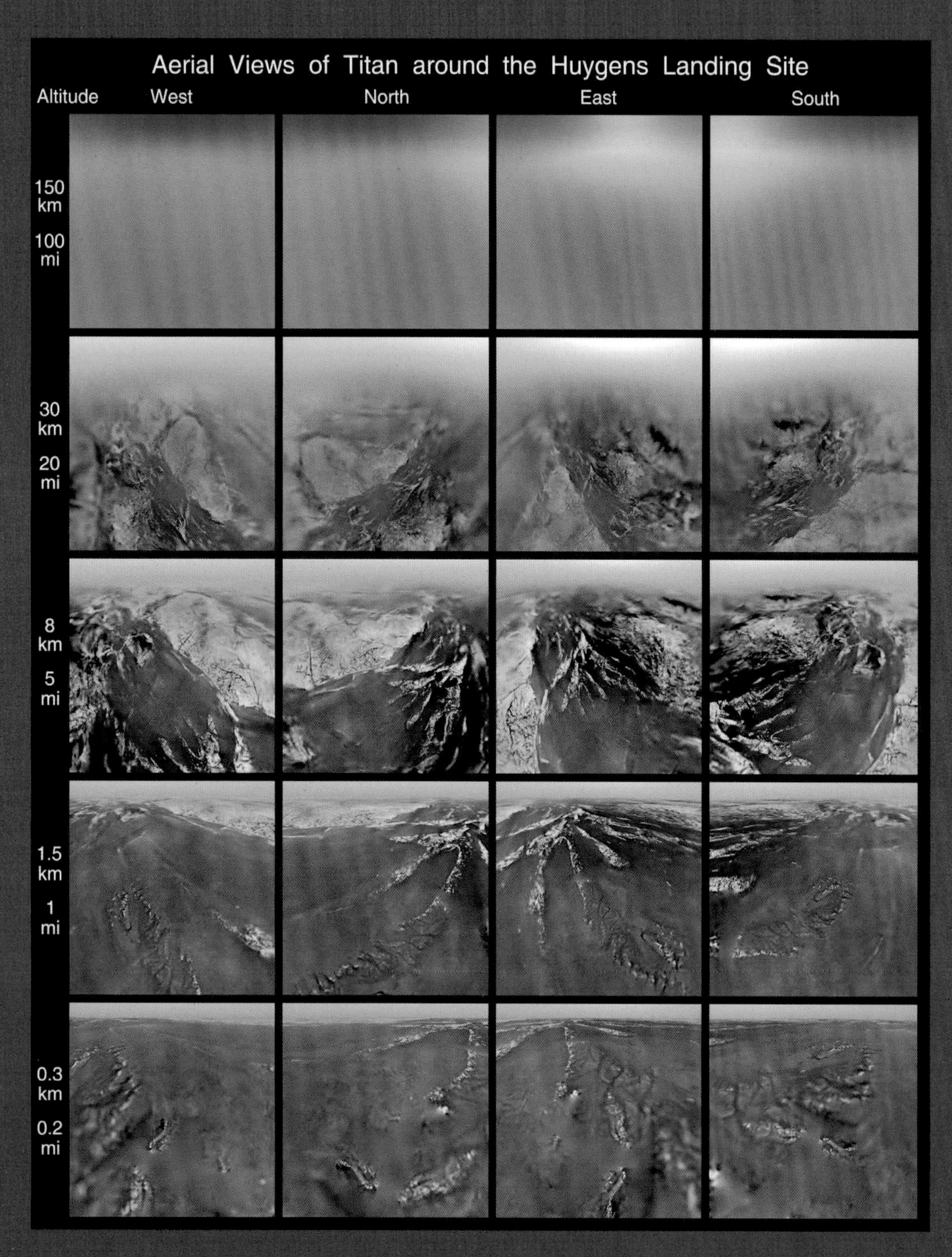

IMAGES TAKEN BY THE HUYGENS PROBE AS IT DESCENDED TO THE SURFACE OF TITAN. CREDIT: ESA.

TITAN IN FRONT OF SATURN.
IMAGE TAKEN BY THE CASSINI SPACECRAFT.
CREDIT: NASA/JPL-CALTECH/SPACE SCIENCE INSTITUTE.

HONOURABLE MENTIONS

PLUTO, WITH COLOUR DIFFERENCES ENHANCED TO HIGHLIGHT CONTRAST. IMAGE TAKEN BY THE NEW HORIZONS SPACECRAFT. CREDIT: NASA/JHUAPL/SWRI.

THE DWARF PLANET/MOON CHARON. IMAGE TAKEN BY THE NEW HORIZONS SPACECRAFT. CREDIT: NASA/APL/SWRI.

CHARON

Pluto's dwarf companion

When Pluto was demoted from planet to dwarf planet in 2006, it caused an uproar.

Children wrote letters to the International Astronomical Union, the group responsible for the reclassification, begging them to reverse their decision. Grown adults refused to adopt the new terminology. Astronomers received death threats. It was a big deal, and there are still plenty of people today who refuse to accept the change.

The decision was ultimately a logical one. New objects about the size of Pluto and sometimes bigger were being discovered all the time, so the International Astronomical Union had to either add those to the official list of planets, or narrow the definition of a planet. They chose the latter, leaving the term 'planet' to the large bodies that clearly dominate the solar system and adding a new designation, 'dwarf planet', to describe objects like Pluto. Ultimately, I'd argue that this did Pluto a favour. Iinstead of being the tiniest and most distant planet, it is now the most famous of the dwarf planets.

Pluto's status as a planet was questioned long before 2006 when its largest moon, Charon, was discovered. At first, scientists studying Pluto thought it was a lot larger than it actually was because they mistook the Pluto–Charon duo for one object. Charon, like Pluto, is a tiny ice world. It is about half the size of Pluto and orbits extremely close to it, at about 20,000 kilometres from the planet, and from afar (as in from Earth, more than five billion kilometres away), the two sort of blurred together. When scientists realised Charon was a separate object, this meant Pluto was smaller and less massive than they'd originally thought.

Charon's status as a moon might also come up for debate soon. Charon and Pluto technically orbit each other, with the centre of gravity lying somewhere in between them. Like a lot of moons described in this book, Charon is tidally locked to Pluto, meaning it always shows Pluto the same face. Unlike other planets, though, Pluto is also tidally locked to Charon. The two always face each other as they go around and around. Because they're so mutually influential, scientists argue that we should call Pluto and Charon a pair of dwarf planets in a binary system rather than a dwarf planet and its moon.

The New Horizons mission is the only one to have studied Pluto up close. The NASA mission launched just a few months before the International Astronomical Union announced that Pluto had been demoted, and the spacecraft reached the Pluto system nine years later, in 2015. It took that long because Pluto is just that far away. New Horizons actually broke the record for the fastest-moving human-made object in existence at the time, moving through space at over 84,000 kilometres per hour.

New Horizons got a good look at both Pluto and Charon, and the data and images it sent back showed two beautiful, intriguing worlds worthy of scientific interest regardless of planetary status. Despite their small size, Pluto and Charon both have huge mountain ranges, suggesting some kind of tectonic activity in the past. They both also have reddish colouration on parts of their surfaces, hinting at some kind of process that scientists don't yet understand.

Because New Horizons was moving so fast when it passed the Pluto system, and because Pluto and its moons are so small and tight-knit, the encounter was brief. The data and images sent home raised just as many questions as they answered. Planet or not, Pluto and its moon/dwarf companion are fascinating little worlds that will hopefully merit another mission.

SATURN'S MOON MIMAS. IMAGE TAKEN BY THE CASSINI SPACECRAFT. CREDIT: NASA/JPL-CALTECH/SPACE SCIENCE INSTITUTE.

MIMAS

The real-life Death Star

Saturn's moon Mimas has two main features of interest.

First, it's the smallest object in the solar system to be gravitationally rounded. Its surface area is about 500 square kilometres, about the same size as Spain. This appears to be about as small as you can get without becoming a lumpy, uneven object like most of the solar system's smaller moons.

And second, perhaps more importantly, it bears an uncanny resemblance to the Death Star.

Funnily enough, when George Lucas wrote the line 'That's no moon ... it's a space station', in 1977's *Star Wars: Episode IV – A New Hope*, Mimas had not yet been observed closely enough to see what it looked like. The first missions to get close-up images of Mimas were NASA's Voyager 1 and 2 in 1980. It's just a wonderful coincidence that the giant space station masquerading as a moon happens to look uncannily like a real moon in our very own solar system.

Mimas looks this way because of a giant crater measuring 130 kilometres across, giving it a diameter a third that of the entire moon. The impact that made this crater almost destroyed Mimas, and there are fractures on the far side of Mimas from the shock that ran all the way through it. The walls of the crater are as high as five kilometres and its central peak rises even higher than that.
This little moon might also hold yet another subsurface ocean, but it would be the first to show no external signs at all. The ocean moons we know of have signs of tectonic activity from their crusts floating on top of a layer of liquid. Mimas shows no such signs, but measurements of its wobble suggest there might be liquid in there somewhere – sort of like how an egg wobbles when you spin it because of the liquid inside. This doesn't necessarily mean that Mimas might be hospitable to life as we know it, but there might just be an interesting new kind of watery world to discover.

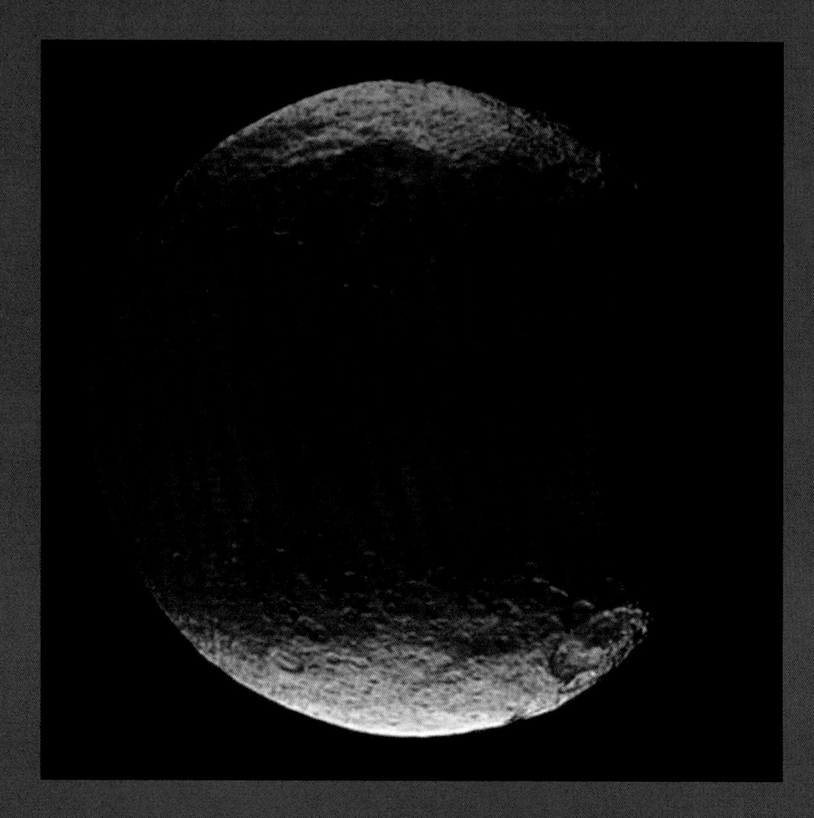

THE TWO SIDES OF IAPETUS. IMAGE TAKEN BY THE CASSINI SPACECRAFT. CREDIT: NASA/JPL/SPACE SCIENCE INSTITUTE.

THE EQUATORIAL RIDGE OF IAPETUS. IMAGE TAKEN BY THE CASSINI SPACECRAFT. CREDIT: NASA/JPL-CALTECH/SSI/EMILY LAKDAWALLA.

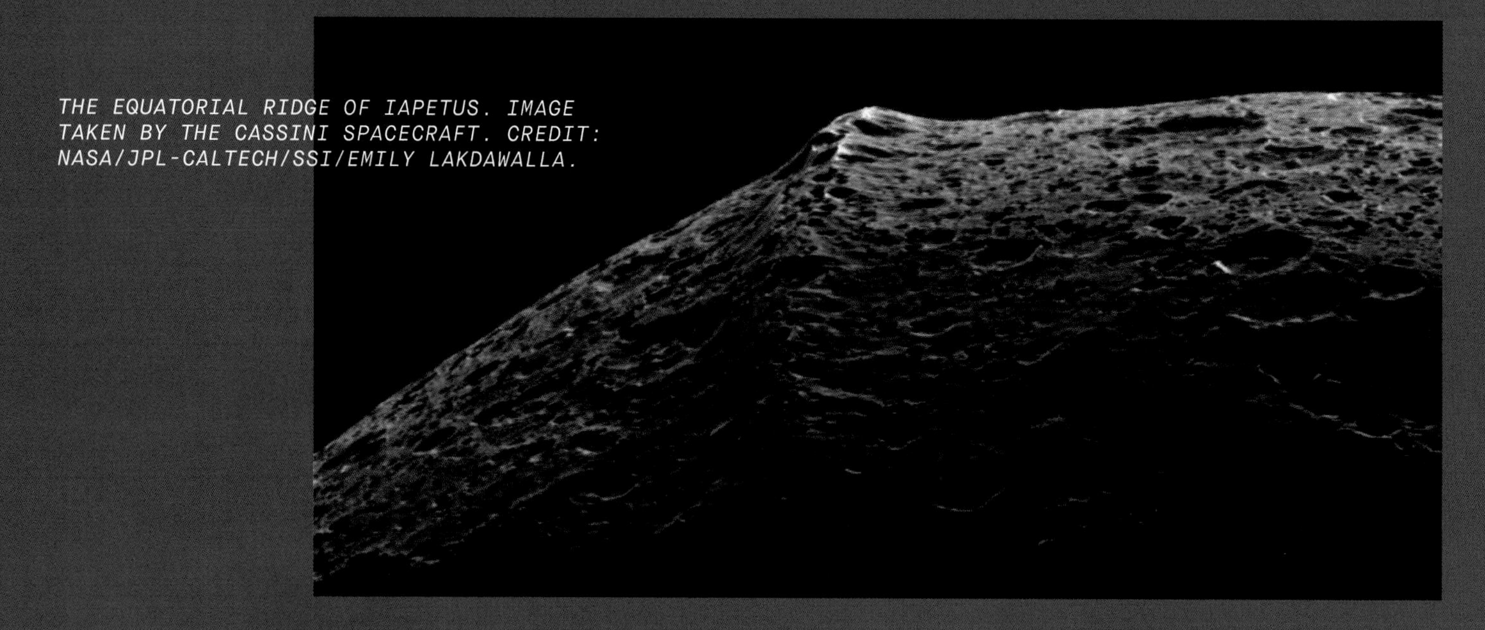

IAPETUS
The two-faced moon

Similar to Mimas, Saturn's third-largest moon Iapetus has a huge impact crater that spans 580 kilometres across (which is 40 per cent of the moon's diameter).

What makes Iapetus really interesting, however, is the stark difference between its two hemispheres. On the side with the crater, Iapetus is bright white with a few dark speckles. On the opposite side, it's almost completely covered in a dark substance – basically, something akin to space dirt.

Scientists first thought that Iapetus was dirty on one side because that was the moon's leading side. It's the side that faces forward as it moves through space, so if there's dirt in its path, that's the side that will pick it up. It's like a car windshield getting littered with bugs. While this theory could explain Iapetus' appearance, it doesn't explain why we don't see this on other moons that are tidally locked to their host planet (rotating at the right speed to always show the same face to the planet) and therefore have a leading side. Iapetus might be getting special treatment by a nearby moon, Phoebe, which is made of darker material and could potentially be spewing some of that material out into space for Iapetus to run into. But we don't have concrete proof.

Another theory is that Iapetus' split appearance reinforces itself due to heat. If, at some point in the past, Iapetus had ploughed into some dark material, that could have set in motion a pattern of heating – the darker side absorbs more heat than the lighter side, and as it warms up it cooks off any remaining ices, making it even darker, absorbing more light, and on and on.

Another puzzling feature of Iapetus is that it has a chain of ten-kilometre-high mountains all the way around its equator, kind of like a rubber ball or a plastic Easter egg. This is a unique feature, and scientists don't really know how it happened. One possibility is that Iapetus used to rotate a lot faster than it does today, and in doing so, it spun some of its mass out towards the equator. Another theory is that it used to have a ring around it, which got dragged down onto the surface and settled into this mountain form.

For now, Iapetus is still a mystery. But as science makes its never-ending march forward, here's hoping we'll eventually crack the case.

NIX

Pluto's chaotic little moon

Pluto's satellite Nix is a truly diminutive moon.

It's shaped like an egg, and only measures about 50 kilometres at its widest. Nix gets an honourable mention because its movement through space is charmingly unpredictable.

Scientists think Pluto's moons were likely formed by a big collision between Pluto and another Kuiper Belt object, much like how Earth's Moon was formed. But because these are all much smaller objects, they didn't gravitationally pull themselves into order the same way the Earth system did. Instead of rotating in an orderly manner like most larger moons do, Nix tumbles chaotically as it orbits Pluto. So, if you somehow found yourself on Nix, you might see the Sun rise in the east and set in the north one day, then do something totally different on another day.

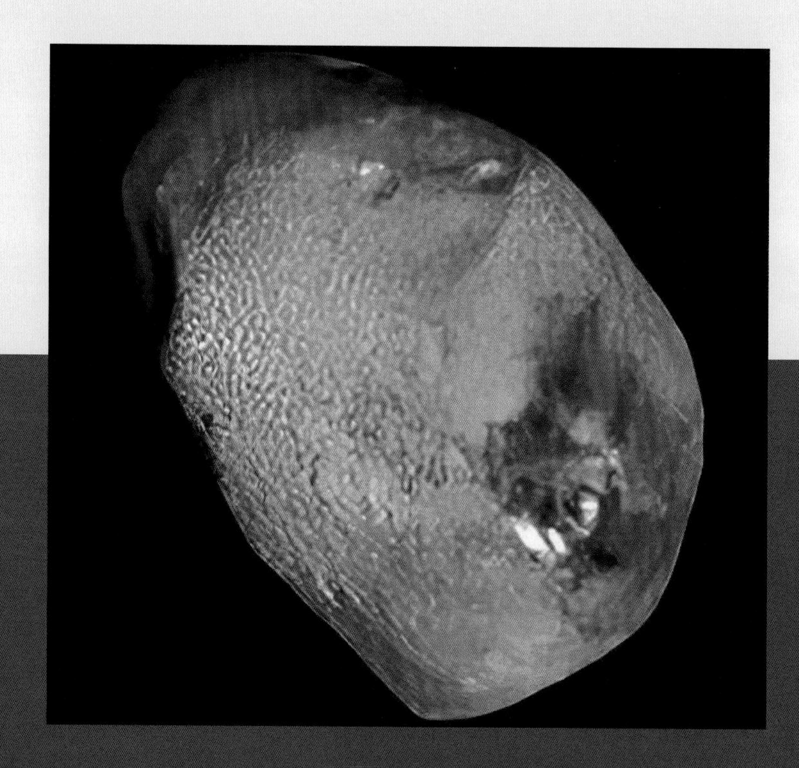

PLUTO'S TINY MOON NIX, IMAGED BY THE NEW HORIZONS SPACECRAFT. CREDIT: NASA/JHU-APL/SWRI.

PAN

The dumpling moon

Topping the charts for cutest moon is Pan, the innermost of Saturn's named moons. (There are other tinier moons without proper names.)

It wins the cute contest for three reasons:

1. It's teeny-weeny, at only 35 kilometres across at its widest point. And, as we all know, small = cute.
2. It looks like a little ravioli, or pierogi, or dumpling, or empanada, or whatever kind of doughy pocket you like. And that's just plain cute.
3. It's known as a ring shepherd because it keeps one of the gaps between Saturn's rings clear of particles. What's cuter than a tiny moon with a job?

The reason Pan has its dumpling shape is actually because of its adorable little job. Because of Pan's orientation as it moves through space, the loose ring material it picks up settles on its surface around its equator. It's also possible that it got this shape when two even smaller, presumably cuter, moonlets crashed into each other and merged, creating a ridge where they combined.

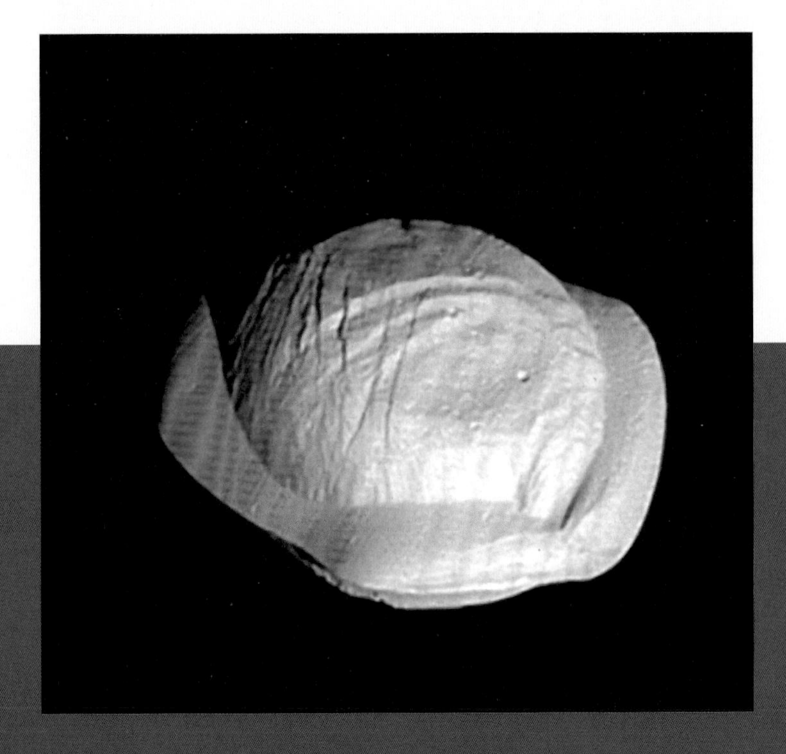

CAPTION: SATURN'S MOON PAN, IMAGED BY THE CASSINI SPACECRAFT. CREDIT: NASA/JPL-CALTECH/SSI.

THE HUBBLE SPACE TELESCOPE'S VIEW OF DIMORPHOS TWO MONTHS AFTER DART'S IMPACT. THE TINY WHITE DOTS CLUSTERED AROUND THE CENTRAL BRIGHT OBJECT ARE BOULDERS THAT WERE THROWN OFF THE ASTEROID MOONLET UPON IMPACT. CREDIT: NASA/ESA/D. JEWITT (UCLA).

DART'S LAST COMPLETE IMAGE OF THE DIMORPHOS ASTEROID BEFORE IMPACT. CREDIT: NASA/JOHNS HOPKINS APL.

DIMORPHOS

The smashed-up asteroid moonlet

Dimorphos is unique in this book in that it's a moonlet of an asteroid, not a planet.

It's technically also an asteroid itself, but because it orbits another larger asteroid, it gets the title of moonlet as well.

Dimorphos is noteworthy because it's the only celestial body in the universe that humans have intentionally moved. In 2022, we smashed a spacecraft into it on purpose to try to change its trajectory around its host asteroid, Didymos. The point of this collision was to test how well an impact like that could divert an asteroid, in case we ever need to use that technique to get an incoming asteroid off a collision course with Earth. The test was a success overall, moving Dimorphos even more off-course than we'd expected – though not enough to accidentally send it on a collision course with Earth, thankfully.

The impact was exciting and gripping, especially because DART (the Double Asteroid Redirection Test spacecraft that did the smashing) had a camera on board that was sending footage back to Earth. The world got to watch live as the spacecraft got closer and closer to Dimorphos until the little asteroid moonlet took up the whole screen, boulders coming into resolution, before the feed cut out upon impact. Very dramatic.

The impact also created a beautiful tail, like you'd normally see coming off a comet. But instead of being ice and dust cooked off the comet's surface by the Sun's energy, it was a tail of dirt from the impact left behind as the asteroid duo made its way around the Sun.

Dimorphos is also a very rudimentary form of celestial object. It's the type of asteroid that scientists call a 'rubble pile', which is pretty much exactly what it sounds like. The entire thing is a collection of rocks loosely held together by gravity. At only 170 metres across, Dimorphos isn't big enough to have gravitationally fused all those rocks together into one solid thing. So it's literally just a pile of rubble floating through space.

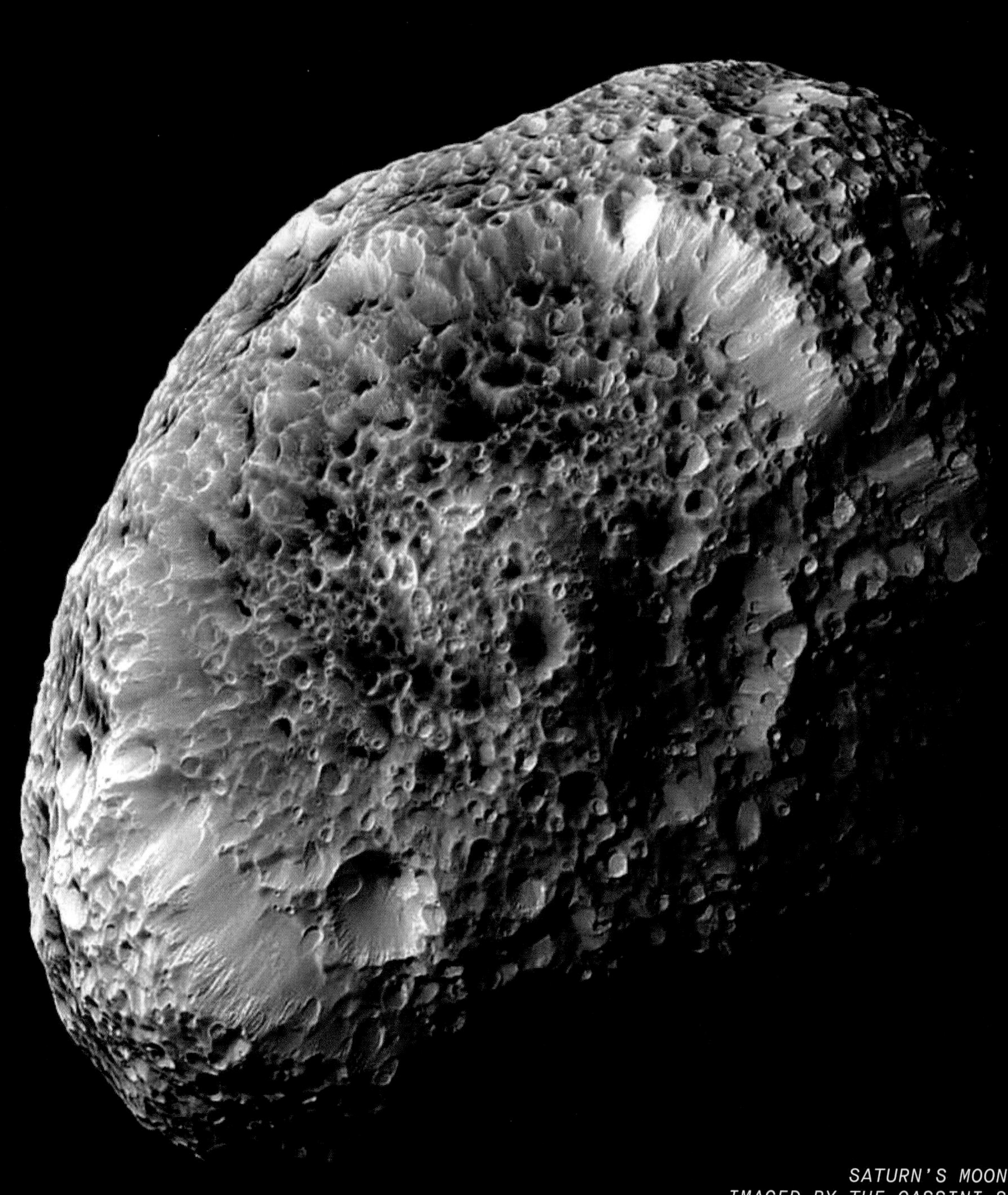

SATURN'S MOON HYPERION,
IMAGED BY THE CASSINI SPACECRAFT.
CREDIT: NASA/ESA/JPL/SSI/CASSINI IMAGING TEAM.

HYPERION

The celestial sponge moon

Another character in Saturn's flock of moons is Hyperion.

Its weirdness is obvious at first glance: it looks like a sponge. A Nestlé Aero candy bar. Perhaps a pumice stone that you'd use to buff dead skin off your heel in the shower.

Hyperion is indeed sponge-like in that about 40 per cent of it is just empty space. It has these pockets of nothingness throughout it. Scientists think it might be porous like this because it formed from smaller objects coming together gravitationally but never getting big enough to fuse into one solid object. Unlike a rocky asteroid, though, Hyperion is mostly made of water ice.

Hyperion is almost big enough to pull itself together into a round object, and is the largest object in the solar system with an irregular shape. It's a bit under 500 kilometres across at its widest, which is bigger than Mimas, the smallest thing in the solar system to have become round through gravitational effects. But because it's so porous, Hyperion doesn't have enough mass for that rounding to happen. And unfortunately, when smaller objects smash into Hyperion they don't usually add to its size; because of its low gravity, impactors tend to blast material off its surface and into space, never to return again.

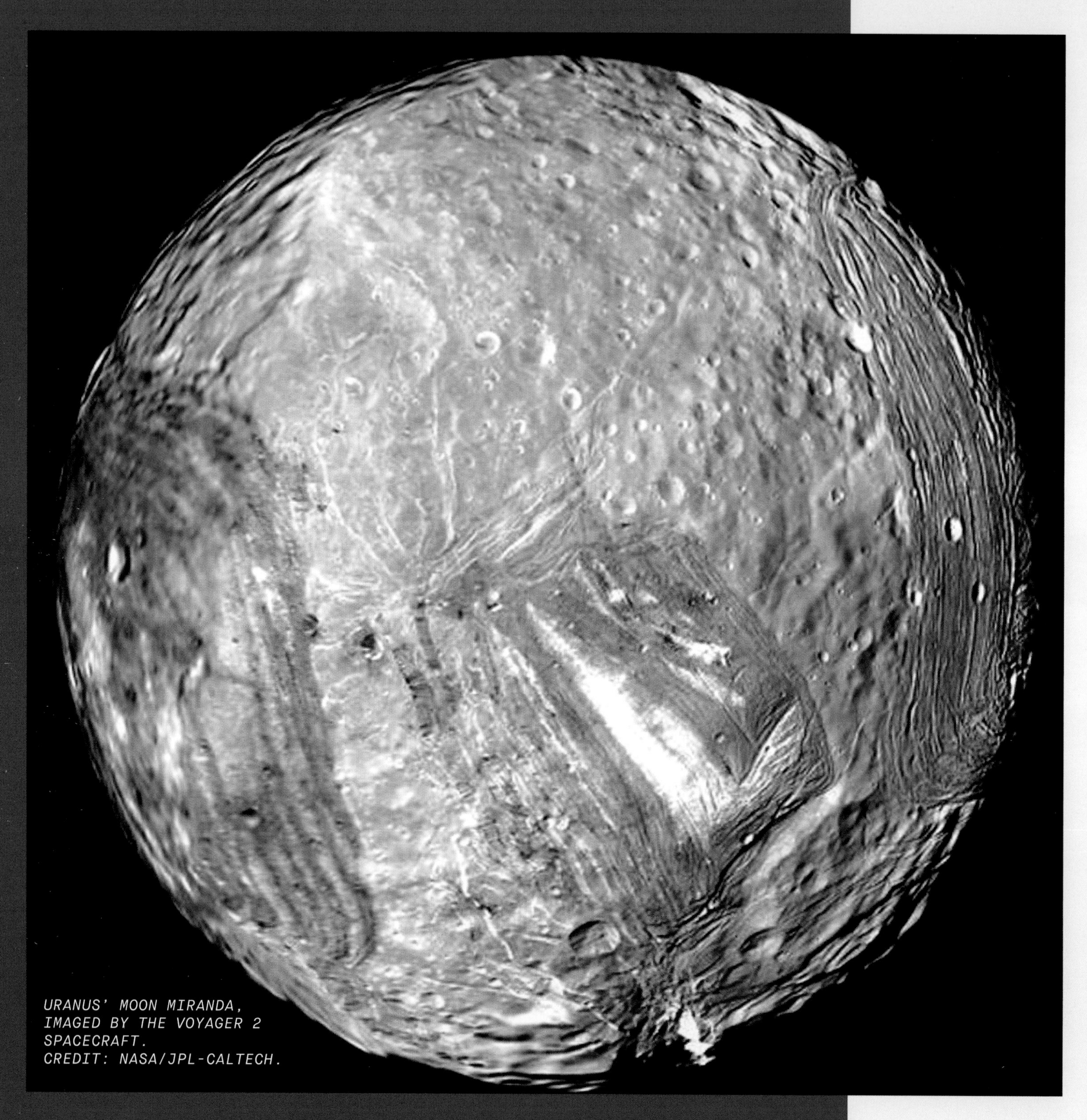

URANUS' MOON MIRANDA, IMAGED BY THE VOYAGER 2 SPACECRAFT. CREDIT: NASA/JPL-CALTECH.

MIRANDA

The Frankenstein moon

So far in this book, we've neglected Uranus' moons, so let's remedy that. Of the icy giant's 27 known moons, Miranda is my favourite.

Miranda is one of the five Uranian moons big enough to have rounded themselves out, but it's far from a perfect sphere. Its surface looks like it's made up of different pieces stuck together, with wildly varying topographies on each section.

At some of the apparent junctions between sections there are huge cliffs, including the highest cliff in the solar system, a staggering 20 kilometres high. With extremely low gravity (the whole moon is only a seventh the diameter of Earth's Moon), if you dropped a rock off the top of this cliff, it would take about ten minutes to hit the ground. Miranda also has enormous canyons, some hundreds of kilometres long and tens of kilometres wide, and reaching 12 times deeper than the Grand Canyon.

As with so many other relatively unexplored moons, scientists still don't know why Miranda is the way it is. It's too small to have internal geological processes causing cliffs and canyons to form, but it's possible that something happened in the early days of its formation. One theory is that after Miranda coalesced, it froze first on the outside and then on the inside as it cooled down; since water expands as it freezes, the inner layers could have then split apart the outer layers that had hardened first, causing all this jaggedness. Another possibility is that Miranda was smashed apart at some point in the past, but not hard enough to eviscerate it. Instead, the impact might have shattered it into small parts that then came back together gravitationally, reassembling into a Frankenstein's monster of a moon.

Miranda may get a visit within the next couple of decades. Every ten years the US National Academy of Sciences brings together planetary scientists to figure out the exploration priorities for the next ten-year period. In 2023, the newest set of decadal priorities included a mission to Uranus to study this mysterious planet and its collection of moons. If NASA follows this advice (which it usually does), a mission to Uranus could launch as soon as 2031 and we'd get to learn a whole lot more about Miranda and its Uranian neighbours.

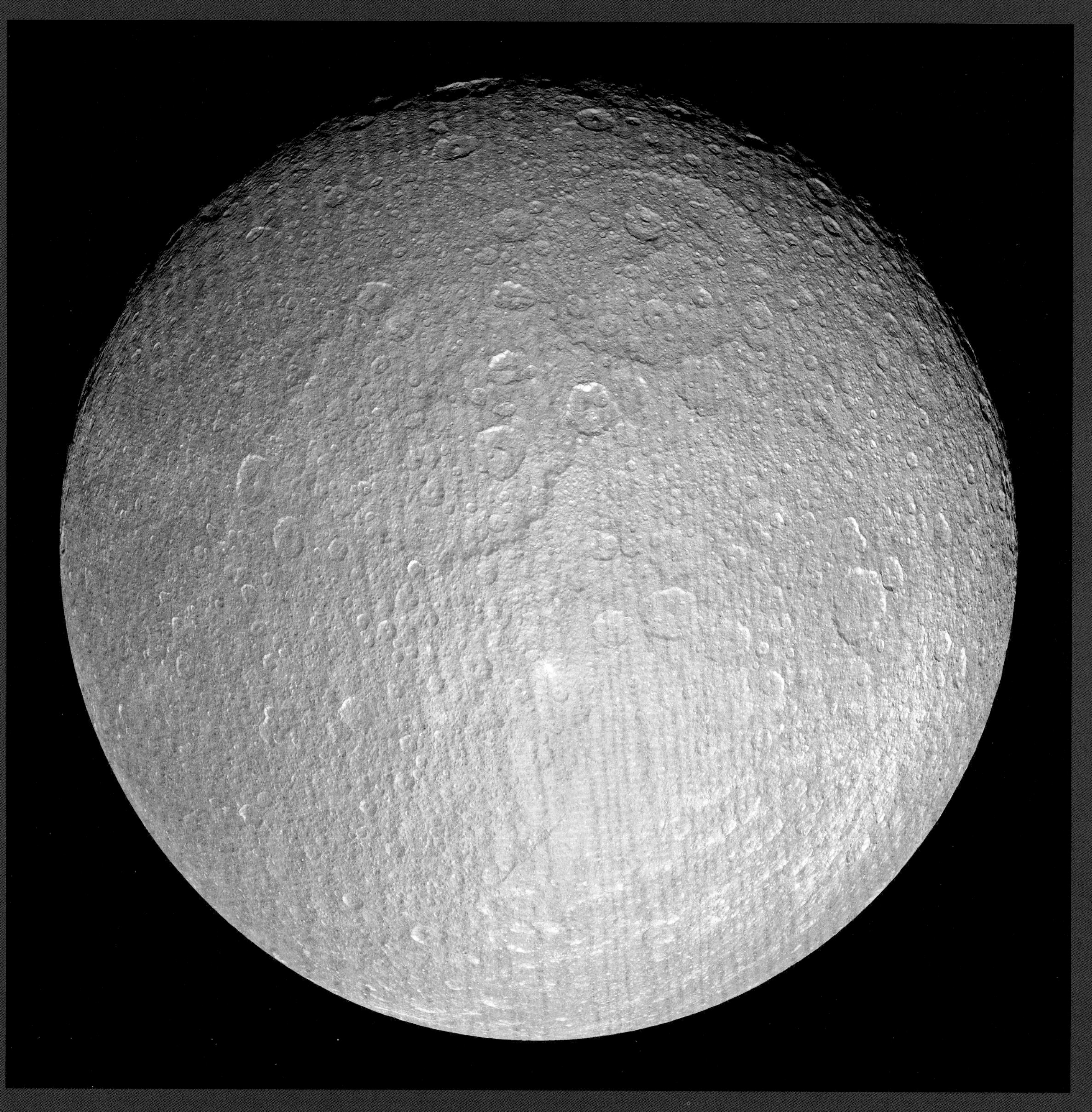

SATURN'S MOON RHEA, IMAGED BY THE CASSINI SPACECRAFT. CREDIT: NASA/JPL-CALTECH/SSI.

RHEA

The ringed moon

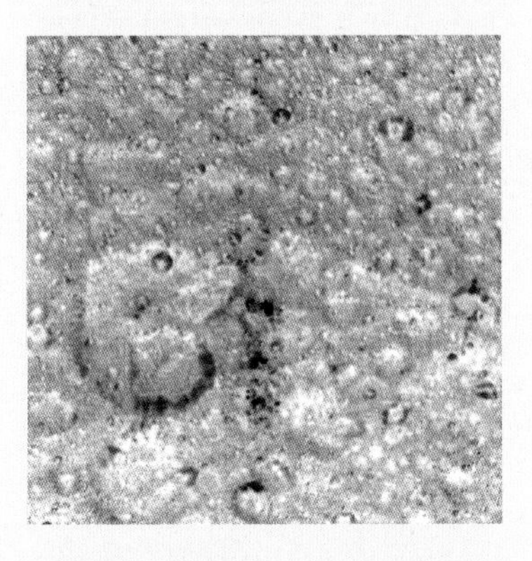

THREE COLOUR-ENHANCED IMAGES FROM THE CASSINI SPACECRAFT SHOWING PARTS OF THE CHAIN OF SPLOTCHES ALONG RHEA'S EQUATOR WHERE FRESH, BLUISH ICE HAS BEEN EXPOSED. CREDIT: NASA/JPL-CALTECH/SSI/LPI.

Rhea, Saturn's second-largest moon, is the only moon we've ever found with a set of rings around it.

The moon itself is similar to a lot of outer solar system moons: made mostly of ice, with no evidence of subsurface water, no tectonic activity, just a bunch of craters on an otherwise smooth surface. But in 2008, the Cassini spacecraft detected a thin band of material orbiting Rhea in the shape of a ring. The ring itself is very small (maybe a metre at its widest), but it is still remarkable since no other moons – not even the big ones – have their own rings.

The evidence for Rhea's ring isn't conclusive yet, since Cassini was only able to detect it indirectly. The spacecraft's instruments were looking at Saturn's magnetic field and found that it was being affected by something around Rhea, an effect that would most likely be explained by the presence of a ring. Cassini was never able to get direct images of the ring, and that spacecraft is now long gone, having intentionally crashed into Saturn to burn up in a dramatic end to its mission.

Another hint that Rhea might have rings – or might have had rings in the past – is that there is a chain of blueish material all around its equator, which could have come from ring material settling onto the surface.

This is reason enough to make Rhea worth taking another look at whenever we next get the chance to visit the Saturn system.

THE GREAT

BEYOND

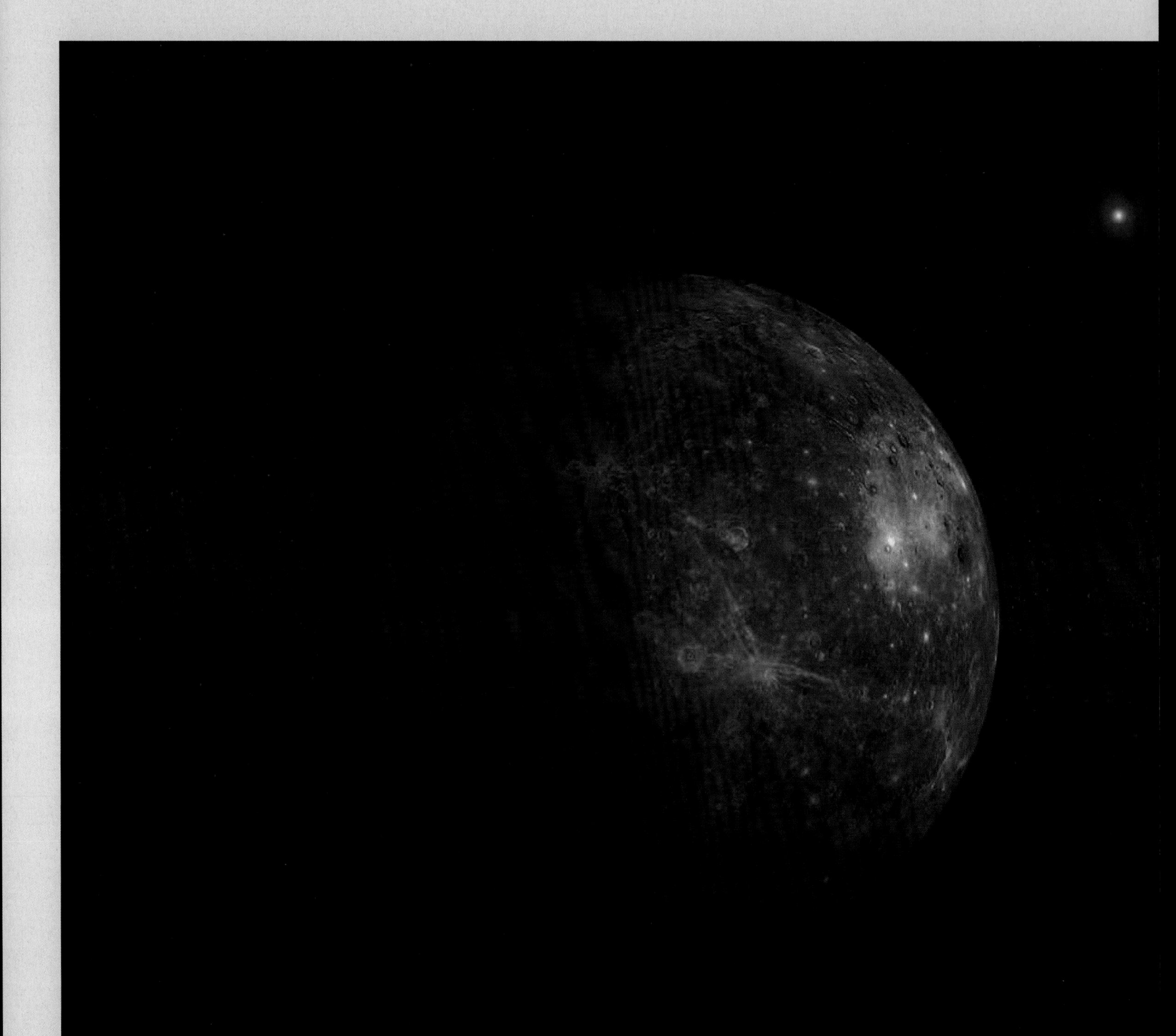

AN ARTIST'S IMPRESSION OF AN EXOMOON. CREDIT: MAURIZIO ARDUINI.

THE MOONS IN THIS BOOK ARE DIVERSE, RANGING FROM FROZEN SNOWBALLS TO LAVA-COVERED HELLSCAPES.

But they all have one thing in common: they're local. All of these moons orbit planets that orbit one star. This is really just one tiny system among the 200 billion trillion (that's 200,000,000, 000,000,000,000,000!) stars thought to exist in the universe. So what else is out there?

The first exoplanet – a planet orbiting another star – was discovered in the mid-1990s. Prior to this, scientists didn't know for sure whether other stars had planets like ours. But once we developed the technology to start looking for exoplanets, the list grew quickly. We have now found more than 5000 exoplanets (a number that will almost definitely be larger by the time you read this), and the data suggest that just about every star out there has at least one planet around it – many are like our Sun and have several planets. With 200 billion trillion stars out there, the number of planets that might exist in the universe is truly mind-boggling. And if those planets are anything like the ones in our solar system, it's likely the number of moons out there is exponentially higher.

To backtrack, it's worth explaining how we know this. Here on Earth, we know the stars as tiny points of light in the night sky. With our most powerful telescopes, they don't get much bigger. Stars are so far away from each other that even our nearest neighbour looks like a tiny dot through the most powerful telescopes we have. So how could we possibly see planets and moons, which are generally thousands of times smaller than their stars?

To add to the size problem, planets tend to orbit pretty close to their stars. We think of the planets as being super far away from each other and from the Sun – after all, Pluto is almost six billion kilometres from our star. But it's all relative – six billion kilometres doesn't look like much from 40 trillion kilometres away (the distance to the nearest star, Proxima Centauri). So when you're looking for planets around another star, you're basically looking directly at that star. And since stars are extremely bright, their glare makes it difficult to see anything near them. It's like trying to spot a bug on the wing mirror of a car a mile away that's shining its high beams at you.

And yet, scientists persevere! Instead of looking directly for exoplanets, they very carefully measure things about the stars themselves to look for indirect evidence of something orbiting them.

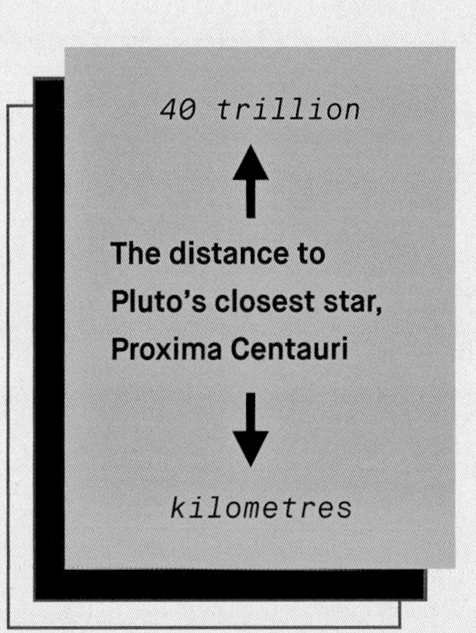

One method for detecting exoplanets is to look for slight motion in the star itself. Even though a star's gravity is much stronger than that of any planets orbiting it, those planets still exert a tiny tug on the star, making it move a bit as a planet goes around it. We've got technology sensitive enough to measure even slight movement in stars, and this can give us evidence that they have one or more planets around them. Another method is to look for slight dips in the brightness of a star caused by planets passing in front of them. Again, these are tiny changes, but we have tools that can sense them.

These techniques are obviously best at detecting whopping big exoplanets that orbit really close to their stars, and we've found a lot of those. But exoplanet-hunting technologies are getting better and better, and detecting smaller exoplanets, some of which seem to have conditions similar to Earth – rocky surfaces that have the right temperatures to support liquid water. And with ever-increasingly sophisticated technologies, exoplanet researchers can now not only spot exoplanets, but also tell their size, composition and atmospheric composition, and even whether they have moons.

Our ability to study exomoons, as they're called, is still extremely rudimentary, since they're much harder to detect than the planets they orbit. But it's a field of study that's growing all the time because scientists recognise that moons are a) awesome, and b) just as likely as planets to be habitable to life.

Exoplanets and exomoons are the most likely place where we'll find signs of alien life. It's really just a numbers game – eight local planets and their moons vs an infinite number of exoplanets and exomoons beyond our solar system. But the way we look for life on exoplanets or exomoons has to be very different from how we do it in our own system. The search for evidence of past or present life within our solar system involves sending spacecraft to visit planets and moons up close, and scratching around on or under the surface for hints of microbes dead or alive. Looking for life in

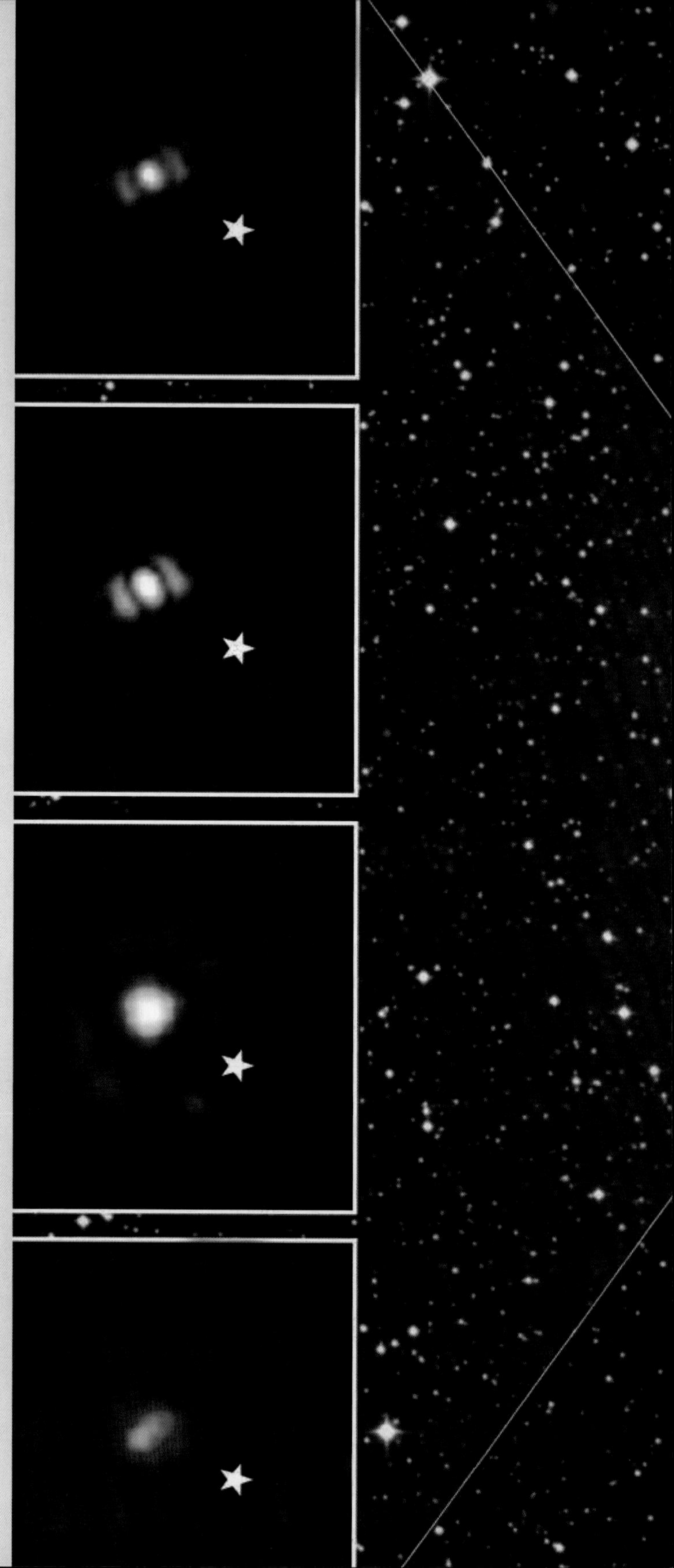

another star system will be less subtle – we'd have to find obvious signs of large-scale life. And that is definitely possible. If you looked at our Sun from a trillion kilometres away and watched the Earth pass in front of it, for example, you could tell there was life here just from measuring the composition of our atmosphere.

The reason for this is that life dramatically affects a planet's atmosphere. On Earth, all the oxygen in our atmosphere (which makes up about 21 per cent of it) comes from living things. Before life emerged on Earth, our atmosphere was mostly carbon dioxide, water vapour and nitrogen. Once living things started consuming energy sources and using them to fuel the processes of life, they also started producing oxygen, among other things. Oxygen isn't an element that can stick around in atmospheres very easily, so it has to be constantly replenished. The same is true of methane – the methane in our atmosphere comes from life forms (much of it from farm animals burping). We do know of non-biological sources of methane, like what we see on Titan, but it's one of those chemicals that raises big flags when spotted in an atmosphere because it has to be produced by something.

These kinds of signs of life are called biosignatures because they're signs of biological activity, but there are also technosignatures: signs of technological activity. From afar, Earth would also be very obviously inhabited because of the technological stuff we've been doing for several centuries now. Cities on Earth's night side are illuminated by lights that can be seen from space. Industrial pollutants in our atmosphere can be detected from vast distances. And the remote communication we do – radio broadcasts, commands sent out to distant spacecraft, cell towers, wi-fi – is all detectable from afar because every one of those forms of communication involves sending out electromagnetic signals, which don't just stop when they're received at the other end. Signal transmissions from Earth have gone out into space, and some can be detected from over a hundred light-years away (since that's how long we've been sending radio signals).

THE JAMES WEBB SPACE TELESCOPE TOOK ITS FIRST DIRECT IMAGE OF AN EXOPLANET – SHOWN HERE IN FOUR DIFFERENT WAVELENGTHS OF LIGHT – IN 2022. CREDIT: NASA/ESA/CSA/A CARTER/UCSC/THE ERS 1386 TEAM/A. PAGAN/STSCI.

SETI

The Search for Extraterrestrial Intelligence (aka SETI) is a very real scientific field that listens for radio signals from other star systems. SETI has been a part of astronomy for centuries, and it really picked up speed as technologies for observing and exploring space improved.

THE VERY LARGE ARRAY IN NEW MEXICO, USA, USES 27 RADIO ANTENNAS WORKING TOGETHER AS A SINGLE RADIO TELESCOPE. THIS ARRAY IS USED, AMONG OTHER THINGS, FOR SETI RESEARCH. CREDIT: NRAO/AUI/NSF.

In the early twentieth century, people were still convinced that there had to be life on Mars because we hadn't had the chance yet to send a spacecraft there. In 1924, there were a few days when Mars would be closer to Earth than it would be for the rest of the century, so the scientific community flagged this as an opportunity to listen for Martian radio transmissions. The United States organised National Radio Silence Day, when every hour for five minutes, all the nation's radios were turned off to listen for signals from Mars without interference. This was an epic undertaking – imagine trying to do anything like that today. But, not surprisingly, it was a futile exercise because nobody on Mars was beaming transmissions out into space for us to pick up.

Nevertheless, SETI projects continue to listen to radio signals from space, and can now target specific star systems that are known to have planets and moons with potentially habitable characteristics.

So the question is: if we found biosignatures or technosignatures on some alien planet or moon, what could we do?

The answer, tragically, is: not much. We could beam a message in their direction, but we wouldn't be able to go visit. Even the closest star, Proxima Centauri, would take about 6300 years to reach using current space travel technology. There's no way we'd be heading over there to see those aliens – which would, more than likely, be microbes with no advice for us on how to peacefully prosper in the cosmos.

That being said, finding life beyond Earth would be extremely exciting regardless of whether or not we could make contact. And it's nice to think that there might be aliens out there, thousands of light-years away, who have spotted Earth and our abundant signs of life and are trying to figure out how to connect with us.

APOLLO 11 ASTRONAUT BUZZ ALDRIN WALKS ON THE SURFACE OF THE MOON ON JULY 20, 1969, IN A PHOTOGRAPH TAKEN BY NEIL ARMSTRONG. CREDIT: HISTORY IN HD

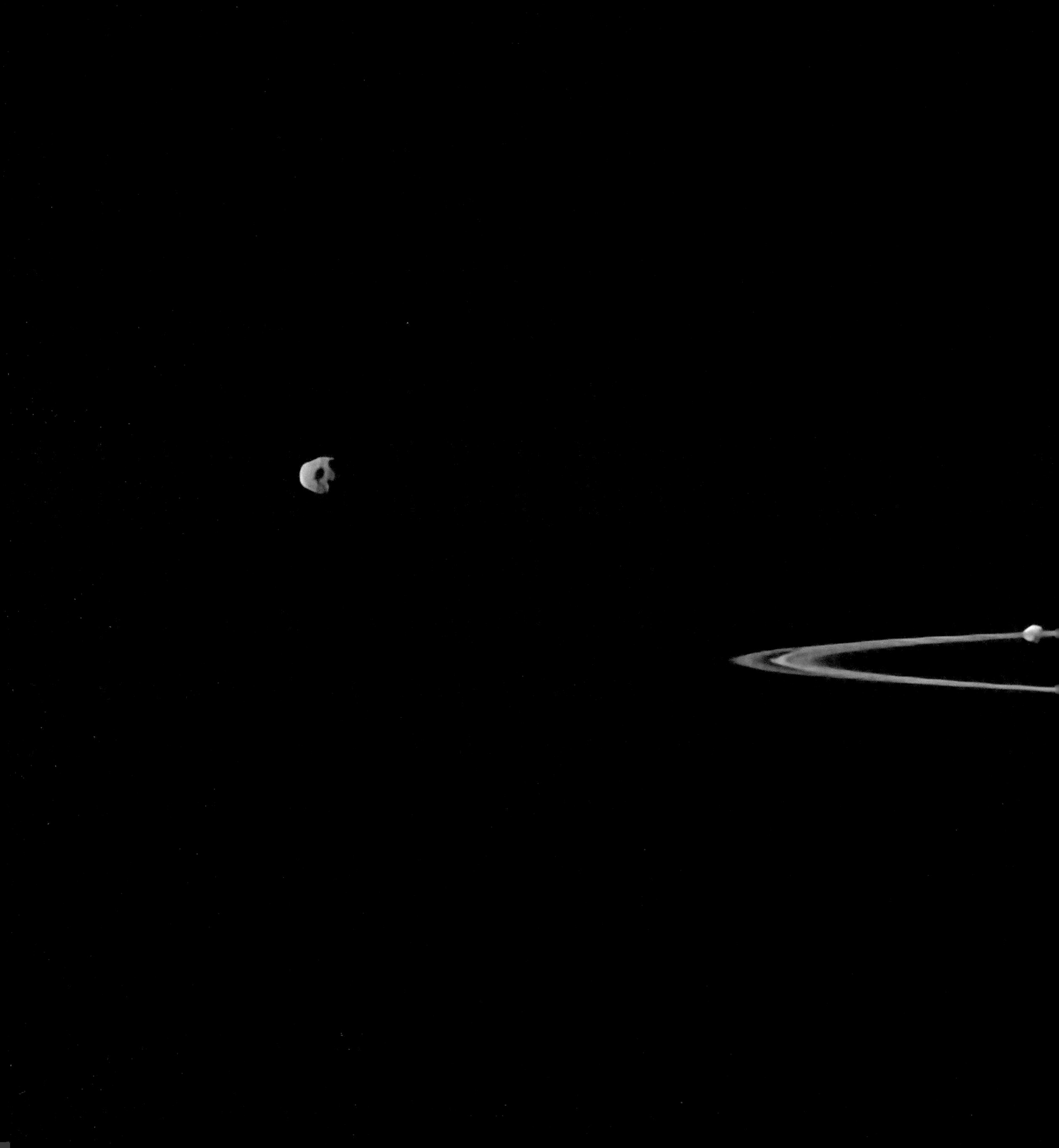

FIVE OF SATURN'S MOONS AND THE EDGE OF ITS RINGS CAPTURED IN ONE IMAGE BY THE CASSINI SPACECRAFT. FROM LEFT TO RIGHT: JANUS, PANDORA, ENCELADUS, RHEA AND MIMAS. CREDIT: NASA/JPL-CALTECH/SSI.

CREDIT: ZOLTAN TASI

CONCLUSION

FOR PEOPLE JUST GETTING INTERESTED IN SPACE, THE SEEMINGLY SMALL STEP FROM LEARNING ABOUT PLANETS TO LEARNING ABOUT MOONS IS A GIANT LEAP.

The universe seems to get a lot bigger and fuller when you realise that around those eight big worlds there are hundreds of fascinating - albeit smaller - new worlds to discover.

Although this book limited its scope to a handful of the coolest moons (at least, in my opinion), there is a lot more to explore. I didn't talk about Dione, Tethys, Titania, Oberon, Umbriel or Ariel. Didn't even bother to mention Dysnomia, Proteus, Nereid, Vanth or Hi'iaka. I'll admit I don't even know anything about Sycorax, Larissa, Galatea, Despina, Namaka or Weywot. The names alone give you some indication of how impressive, strange and interesting these moons could be.

And the list keeps growing. I'll bet that within a few years of this book's publication, there will be a whole new family of exomoons to talk about – and even new moons in our own solar system. While I was editing the first draft, 12 more small moons were discovered around Jupiter. And I'm sure that new, exciting discoveries will continue to be made about many of the moons covered in this book. With every discovery we make about the moons of our solar system, they rise up the ranks of places we want to explore. Space agencies around the world are increasingly setting their sights on moons, and missions to Europa, Callisto, Ganymede and Titan are already in the works. Despite all the exploration that's yielded the knowledge I've shared in this book, the heyday of moon science is yet to come. It might not be that long before we find evidence of life beyond Earth, and that life may well be found on a moon. It could be in another star system, or it could be close enough to home for us to study it extensively and make giant leaps in our understanding of what it means to be a living thing.

Even if moons don't wind up being the place where we find life, they still have so much to teach us about the staggering diversity that the universe is capable of producing. It's awe-inspiring to think that a few basic laws of physics can wind up producing all these weird and wonderful worlds, and that we – a bunch of naked apes – have the ability to visit them up close, study them from afar, deduce all kinds of things about what's happening under their surfaces, and even make plans to live on them. That combination of nature's ability to be awesome and our ability to figure out and appreciate that awesomeness is what makes space exploration so extremely cool. And it's wonderful to know that as knowledge builds on itself, our technologies continue to advance.

THE BEST IS YET TO COME.

APOLLO 17 ASTRONAUT HARRISON SCHMITT WITH THE LUNAR ROVING VEHICLE AT SHORTY CRATER, PHOTOGRAPHED BY ASTRONAUT GENE CERNAN. CREDIT: APOLLO 17 CREW/NASA.

Kate Howells is a science communicator with a passion for how weird and wonderful the universe can be. She works as a public education specialist for The Planetary Society, a space advocacy organization co-founded by Carl Sagan, and serves on the board of directors of the science outreach group Royal City Science in her hometown of Guelph, Ontario, Canada. Kate's mission is to encourage people to be curious about science and nature and to share in the enjoyment and appreciation that brings. Kate is also the author of *Space is Cool as Fuck.*